S0-BVP-252

CONSTRUCTION AND USE OF
ATOMIC AND MOLECULAR MODELS

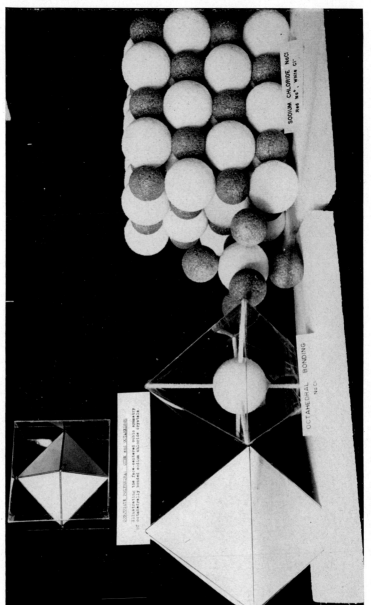

Frontispiece Octahedral coordination (left) of sodium chloride type crystal packing model (right), and relation between

CONSTRUCTION AND USE
OF ATOMIC AND
MOLECULAR MODELS

by

H. BASSOW

1966
THE QUEEN'S AWARD
TO INDUSTRY 1966

PERGAMON PRESS

OXFORD · LONDON · EDINBURGH · NEW YORK
TORONTO · SYDNEY · PARIS · BRAUNSCHWEIG

Pergamon Press Ltd., Headington Hill Hall, Oxford
4 & 5 Fitzroy Square, London W.1
Pergamon Press (Scotland) Ltd., 2 & 3 Teviot Place, Edinburgh 1
Pergamon Press Inc., 44–01 21st Street, Long Island City, New York 11101
Pergamon of Canada Ltd., 207 Queen's Quay West, Toronto 1
Pergamon Press (Aust.) Pty. Ltd., 19a Boundary Street, Rushcutters Bay,
N.S.W. 2011, Australia
Pergamon Press S.A.R.L., 24 rue des Écoles, Paris 5ᵉ
Vieweg & Sohn GmbH, Burgplatz 1, Braunschweig

Printed in Great Britain by A. Wheaton & Co., Exeter

CONTENTS

ACKNOWLEDGEMENTS

THE author wishes to express his appreciation to Thomas R. P. Gibb, Jr., Professor of Chemistry, and Director of Sponsored Research, at Tufts University, who first introduced and aided him in the construction and use of crystal models from styrofoam spheres.

Thanks are also due to Hy Ruchlis, former physics teacher, and presently Director of Harcourt, Brace's Special Projects Department, for his encouragement in the development of simple methods of constructing models from styrofoam.

Finally, the author is grateful to the staff of the Chemical Education Material Study (CHEM Study), the U.S. National Science Foundation sponsored chemistry curriculum improvement project, for its very special contribution to chemical education, encouraging student and teacher alike to build and use the types of models described in this book. Keith McNab, chemistry teacher, and former member of the CHEM Study staff, was especially helpful in this connection.

NATURE OF A MODEL:
THE BLACK BOX ANALOGY

THIS is a book about scientific models, particularly the construction and use of models of different chemical substances. We might very well begin by considering just what such a model is anyway, or—even better—the scientific meaning and use of the term "model".

The dictionary defines "model" as ". . . a representation to show the construction, or serve as a copy, of something". The problem for the scientist is that he does not know what the "something" looks like, and therefore cannot copy it. Thus the scientists' model is not a model in the dictionary sense, since it clearly can NOT be a copy of the real thing.

As we try to understand just what a scientific model is, it may help to imagine you have been handed a sealed box, and asked to construct a mental picture of what an object contained in the box might be like. It would help even more to have a friend prepare such a box for you—without identifying the object he has placed inside—so that you can have the actual experience of trying to evolve an idea of what it might look like.

Suppose, for example, that as you tilt the box gently from side to side you hear a distinct rolling sound. "Aha," you say, "it's a marble! ! " What you mean, of course, is that it rolls like a marble would, so that it could be LIKE a marble at least in this respect. It could also be like a golf ball, or a ball bearing, or perhaps a large wooden bead, and this is why it is better to say it is LIKE a marble instead of it IS a marble. But wait a minute—a pencil would roll, too, in one direction! Perhaps the object is like a pencil. How could you tell? "Why," you say, "tilt the box in the OTHER direction of course!" A rolling sound then would indicate a round, marble-like object, while a sliding sound might suggest the pencil-like possibility. Let us assume, in our example, that the sound is once again of something rolling.

Notice what a remarkable thing you have done here. First, you performed an experiment: tilting the box. You made an observation: you heard a rolling sound. This led you to evolve two mental models for the object in the box: marble-like or pencil-like. These models did three things: (1) they explained your initial observation, since either marble or pencil would roll in one direction; (2) they suggested further experimentation (i.e. tilting the box in the other direction); and (3) they predicted the possible results of such an experiment (either a rolling, or a sliding sound). Finally, you perform this second experiment, and on the basis of further observation (more rolling sound), you choose the marble picture.

This is not bad, and is completely analogous to what scientists do. The point, of course, is that the marble is a MODEL because it would behave as the object did in the situations described above. The fact that a golf ball, ball bearing, wooden bead, or indeed any other spherical object, would behave in a similar manner, only serves to emphasize that the object is in some ways LIKE a marble, but not necessarily a marble after all.

Thus the statement that a marble is a good model for this object brings home the meaning of the term "model", as we use it in science. Never forget that any model this book may help you to build is NOT necessarily an enlarged copy of anything, but simply a scientific model of it. And if such a model explains and predicts some of the behavior of the actual thing, this is all we have a right to expect.

THE ATOMIC MODEL OF MATTER

Introduction

INSTRUCTIONS FOR PREPARING MODELS NEEDED IN UNIT 2

The serious reader is advised to purchase a student organic molecular models ball-and-stick kit, available at most chemical supply houses* at about $5. This consists of wooden spheres of various colors, in which holes have been drilled. The kit also contains dowel-rod connectors, as well as springs useful in representing the organic chemists' "double bonds" that are taken up in later units. This kit will prove invaluable with Units 3, 5 and 6 of this book.

Readers who do not wish to purchase the kit described above may buy quantities of inexpensive styrofoam spheres* and prepare them himself at a fraction of the cost of the ball-and-stick kit. A minimum order of such spheres, which would serve Units 3, 5 and 6, as well as this unit, follows:

No. of spheres	Final color	Diam. (in.)	No. of holes
18	white	1	1
8	black	$1\frac{1}{2}$	4
6	red	$1\frac{1}{2}$	3
4	light green	$1\frac{1}{2}$	1
3	blue	$1\frac{1}{2}$	3

* Sources of supply are listed in the appendix at the end of the book.

Holes may be made by simply pushing round toothpicks half-way into the sphere, directed towards its center. For this unit, where more than one hole is called for, simply make the holes as far away from each other as possible. More accurate methods will be detailed, in later units, as they are needed.

Spheres are usually white when purchased, but may be painted first with white latex-base paint. When dry, a second coat of enamel, of the proper color, may be added, to give a hard, shiny finish. Or, the raw sphere may simply be given a first coat of the proper color latex paint, making a second coat unnecessary. Readers using styrofoam spheres may join them with ordinary toothpicks.

Elements

Scientists have long accepted the idea that all matter is made of one or more of some 100-odd fundamental materials which are called **elements**. If something contains only one kind of element, it is known as an **elementary substance**, or element. Common examples are iron, aluminum, silver, copper, oxygen, etc. It is true that there are many more substances than there are elements. This can be explained by assuming that these elements may be mixed or joined together, thereby forming new substances, in much the same way that our alphabet letters can be rearranged to form new words. Just as our 26-letter alphabet can be used to make millions of different words, so the 100-odd elements can form millions of different substances.

Any substance may be identified by its properties. Thus iron is solid, heavy, lustrous, tough—a typical metal; while oxygen is a tasteless, odorless, colorless, invisible gas needed for respiration. Even though some elements have similar properties, you will always find differences if you look hard enough. Iron, aluminum, copper and silver are all metals, and are therefore similar in many ways. Aluminum is least dense (i.e. a given volume of it is lighter than the same volume of the other metals), while silver is the densest of the four. Copper differs in color from the rest, and copper and silver are better conductors of electricity than the others.

A given sample of an element, when pure (free of all dirt and other foreign matter), has a particular set of properties unique to it and no other element. Thus, a piece of pure silver, no matter where or when or how it was mined, has the same properties as all other pieces of pure silver. These properties in turn (some or all of them) are different from the properties of samples of other elements. The only exception to this fact of nature is that most elements are found to have two or more forms, called **isotopes**, which differ from each other only in weight. This need not concern us here. It is also found that certain elements can exist in what is known as different **allotropic** forms, also of no concern here. You may wish to consult the references at the end of the manual for further information on these exceptions.

Introduction to the Atomic Theory

In order to explain the observed facts about the properties of the elements, as well as many other things, the English scientist Dalton in 1808 proposed an ingenious extension of the ancient idea that matter is made of tiny sub-microscopic (too small to be seen even through a microscope) particles called **atoms**. All the atoms of a given element, said Dalton, are exactly alike in size, shape, weight, color and all other properties. Atoms of other elements, however, have different sets of properties. The fact that this assumption has been slightly modified to account for isotopes does not detract from its usefulness to us here.

By means of colored wooden or styrofoam spheres in your kit, you can see for yourself just how useful this first assumption of the so-called **Atomic Theory** can be. Let each of the spheres represent one atom of a particular element. Five differently colored spheres, therefore, represent atoms of five different elements. Later on, we shall suggest a particular color represent a particular element. For the present, let them represent any five different elements.

PROPERTIES OF ELEMENTS

By means of these colored spheres, you can easily show why every sample of any given pure element is just like every other sample. To do this, divide the eighteen equal-sized white spheres into three groups of six each. These now represent three different samples of what we may call element A. Notice that each group resembles the others in color, size and weight. If we now transfer a sphere from one group to any of the others there will be differences in the size and weight of each sample, but the color of all three will still be the same.

A weighable, visible amount of an element would contain about 60,000,000,000,000,000,000,000,000 atoms instead of six, so it is certain that different samples of a given element would contain different numbers of atoms. They would therefore be expected to have different weights and sizes. If you imagine that color represents all other properties, you can see why the different samples of a given element are otherwise identical.

If you now compare one of these white sphere groups (element A) with one made of eight equal-sized black spheres, it is obvious that they are NOT of the same color. If you imagine that the black spheres represent a sample of what we may call element B, you can see that the different properties of the different elements are explainable in terms of the different properties of the atoms themselves.

REASONING BY ANALOGY

Consider for a moment what a simple, yet remarkable thing you have just done. For one thing, you have accepted the theory that all matter is made of tiny invisible particles called atoms. Then you have represented these atoms you cannot see by wooden spheres which you can see and manipulate. Thus, you have made use of a mechanical model (the spheres) to represent things we can only imagine (the atoms). When you then used groups of similar spheres to represent groups of similar atoms, you were

reasoning by analogy. You could say, in effect, that if the three groups of white spheres were identical in properties because they were all white (i.e. color is analogous to properties), then assuming also that the spheres represent atoms (spheres are analogous to atoms), it follows that what holds true for different groups of the same kind of spheres will also be true for different groups of the same kind of atoms.

This procedure is known as reasoning by analogy. Although it often calls for stretching the imagination somewhat, perhaps you can already see why it is justified. Notice that by means of our mechanical model analogy, we were able to understand and explain something that is actually found to be true in nature: namely, that different samples of the same element are alike in their properties, but differ in these properties from samples of the other elements.

As you proceed with this manual, you will do this kind of thing many times over. It is not surprising, therefore, that such models and procedures are widely used in science, for they are very useful in helping us to understand the world around us. Do not forget, however, that no matter how perfectly a model explains the things we observe to be true, the model was invented by man and is only a model. You should not think for a moment that atoms actually look like the spheres. Although no one knows just what atoms do look like, we know they do not look like these spheres. It is simply convenient for our purposes to assume they are analogous to each other—there is quite a difference.

Elements and Compounds

Let us continue this very effective reasoning by analogy. We can use it and our spheres to see how a limited number of elements might be able to form the millions of different substances which have been found to exist in nature, as well as the many others man has learned to make. All of these different materials are either **pure substances** or **mixtures** of two or more pure substances.

A **pure substance** is just that—something made of one kind of material, and free from all other materials. Any element, providing it contains no impurities, is one type of pure substance. Elements, however, can combine chemically with one another to form completely new substances, with new properties, which are also pure. These are called compound pure substances—**compounds**, for short. There are several ways to tell if two or more elements have so combined. Perhaps the simplest way is to see whether or not they have lost their original properties. If they have, then we know they have chemically combined to form a compound.

Perhaps the most familiar example of this is the rusting of iron. Here, the element iron slowly combines with the element oxygen, present in the air, to form a new substance, the rust. How do we know it is new? It has properties completely different from either of the elements which formed it. Iron is a hard, tough, lustrous solid, which is attracted by a magnet. Oxygen is a colorless, odorless, invisible gas. The rust they form is a reddish-brown, crumbly solid, with no lustre and no magnetic attraction. These properties are obviously new, and hence we may conclude that the rust is a new substance. Since it is compounded of more than one element, we call it a **compound**.

Sometimes the formation of a compound is much more spectacular than this. You may have noticed that a photographic flash bulb initially contains many thin strands of metal inside the bulb. After the flash, these strands are replaced by a dull white powder not at all like the metal strands. In this case, the metal—an element called magnesium—has chemically combined with the oxygen gas inside the bulb to form the white powder. Since this powder resembles neither of the elements initially present, we know that it, too, must be a compound. This time, the formation of the compound was accompanied by an obvious energy change, in the form of the heat and light of the flash. As a matter of fact, it has been found that every change involving the formation of one or more new substances is accompanied by an energy change. Often, as in the case of the rusting of iron, it is not always an obvious one.

HOW ELEMENTS FORM COMPOUNDS

Let us use our spheres to help us see how the atomic theory can explain the formation of compounds. Again, let the spheres represent atoms. Since we also want to represent the joining together of these atoms, we will use toothpicks or longer dowel rods to represent the links, or **bonds**, which hold the elements—and therefore their atoms—together in their compounds. As an example, push a toothpick or dowel rod into one of the white spheres, twisting as you push. Now push a light green sphere onto the free end of the toothpick or dowel. If we let the white sphere represent one atom of what we may call element A, and let the green sphere stand for one atom of another element, B, then the resulting unit of the white and green spheres joined by the dowel (by the analogy we have been using) would represent the smallest possible unit of a compound of these two elements. Indeed, the unit could not be made any smaller without again becoming separate atoms of A and B.

This smallest possible unit of a compound is called a **molecule**, and our white and green model might reasonably be called one molecule of the compound AB. Again, as in the case of atoms making up a visible amount of an element, you would have to imagine the joining process you have just completed as taking place billions of times, between billions of each kind of atom, before there would be enough AB molecules to be seen.

BUILDING COMPOUNDS FROM ELEMENTS

We have already made an analogy between the way in which the relatively few letters of the alphabet can be used to form many different words and the way in which relatively few elements might combine to form many different compounds. Let us investigate this by the use of our colored wooden spheres.

We have already formed one "molecule" of a compound using two different "atoms" representing two different elements—

white and green. If, for the present, we allow only two-atom molecules (i.e. molecules made of only one atom of each element), then there is only one such molecule that the two elements can form. This is the one you have already made. Since the compound consists of many such molecules, it follows that there would be only one possible compound of this type formed from only two different elements.

Now let us try forming such two-atom molecules from, say, four different elements (four different kinds of atoms). Consequently, start with four groups of spheres: four white, four green, four black, and four red—a total of sixteen spheres. With charcoal write the number "1" on each white sphere. Then, using white chalk, mark each green sphere with the number "2". Similarly, write a "3" on each black sphere, and a "4" on each red one. These numbers, which can easily be chalked on, and later rubbed off the spheres, will help you to identify the different possible "molecules" which you can make from them.

Join a different color sphere to each end of as many toothpicks or dowel rods as you need, exactly as you did before, until you have made as many different two-atom molecules as you can from these four different groups of spheres. Notice that once you have joined a white "1" and a green "2" to form what we may call a 1–2 molecule, you may not join another green to another white sphere, for this would result in a second molecule that is the same as the first. You will find that you can make six different molecules, which we can identify by means of the chalked-on numbers as 1–2, 1–3, 1–4, 2–3, 2–4 and 3–4.

Although you may not have enough spheres of each color for you actually to build molecules from more than four groups in this way, you can build them in your imagination by simply assuming that you have a fifth group of spheres of still another color, each of which has the number "5" chalked on. How many different two-atom molecules could you now make? This time you will find the number to be ten. Here are the possibilities, identified as before, and arranged in an orderly way:

```
1–2
1–3    2–3
1–4    2–4    3–4
1–5    2–5    3–5    4–5
```

Now try adding a sixth group, each numbered with a "6". You may be surprised to find that the number of possible molecules has jumped to fifteen. Here they are:

```
1–2
1–3    2–3
1–4    2–4    3–4
1–5    2–5    3–5    4–5
1–6    2–6    3–6    4–6    5–6
```

Adding a seventh group will make the number of possibilities jump to 21. With ten different groups, the number becomes very large. If you like mathematics, you might try to derive a formula that will give the number of possible molecules if you know the number of different elements available. Do not be discouraged if you cannot find one that fits all cases.

What is the point of all this? Remember that our spheres represent atoms of different elements, and the joined combinations stand for molecules of their resulting compounds. By our analogy, six different compounds should be possible from four elements, ten different ones from five elements, and so on. This is with the limitation that we may make only two-atom molecules. The number of possibilities would increase markedly if we allowed more complicated molecules to be built. The model has made its point. It has explained the existence of a very large number of compounds, even though the number of elements from which they are made is strictly limited.

The Weight Laws

Even before Dalton proposed his atomic theory, certain facts connected with the weights of substances before and after their chemical combination were always found to hold true. Whenever any such facts seem always to be true in case after case, with never an exception, they are assumed to be true in all cases. These facts are then summarized in a statement which becomes known as a **universal law of nature**. We shall consider here two such laws, and then go on to see how neatly the atomic theory is able to explain them.

CONSERVATION OF MATTER

The first such weight law to be recognized and stated was concerned with the weights of two or more substances before and after they had chemically combined. In every such case, the total weight of the substances before and after combination was found to be the same. In other words, matter has weight, the amount of matter remained the same—or, more scientifically stated, was conserved. Hence the name, **conservation of mass**. For example, when iron rusts, the weight of the iron plus the weight of the oxygen it combines with equals the weight of the rust formed.

To understand why this law should hold, consider what the atomic theory says happens in terms of atoms during a chemical combination. Recall our combining of white and green "atoms" into molecules. We simply joined them together using dowel rods, or toothpicks, as bonds. If you stop to realize that the bond between any two atoms is some kind of attractive force, and not a piece of wood with weight, it is apparent that the dowel rods, or toothpicks, must be considered weightless if they are to be analogous to these bonds.

Therefore, the only thing that happens during combination is that white and green atoms are put together instead of being kept apart. It is immediately obvious that the weight of a white sphere plus the weight of a green sphere adds up to the same thing whether they are apart or stuck together. By the same token, the weight of a billion white plus a billion green spheres remains the same whether they are separate or joined. Thus, the atomic theory simply and easily explains the law of conservation of matter.

DEFINITE COMPOSITION

The second so-called weight law of interest to us states that every sample of a given pure compound is always made of the same elements in the same proportions by weight. This is known as the **law of definite composition**. Using our flashbulb example, if the white powder found after the flash is analyzed, it is found to contain the elements magnesium and oxygen in a weight ratio of 3 to 2. That is, a 5-gram sample would contain 3 grams of magnesium and 2 grams of oxygen. A 10-gram sample would be 6 grams magnesium and 4 grams oxygen. Five pounds of the stuff would be 3 pounds magnesium and 2 pounds oxygen, and so on. This same analysis would hold for powder from any used flashbulb, or, indeed for any other sample of the powder even if it was made in some other way. As long as it is a sample of this pure white powder, it would always contain 3 parts by weight of magnesium and 2 parts by weight of oxygen.

Pure water, when analyzed, is found always to contain the elements hydrogen and oxygen in a 1 to 8 ratio by weight. Nine pounds of water would therefore contain 1 pound of hydrogen and 8 pounds of oxygen. An 18-pound sample of it would contain 2 pounds of hydrogen and 16 pounds of oxygen, and so on.

Probably the most important postulate of Dalton's atomic theory was the one in which he assigned relative weights to the different atoms. It was important because with it he was able to explain all of the above. These weights became known as **atomic weights**, and although we no longer accept Dalton's values, the concept remains a very useful one.

Obviously, we cannot weigh an atom on a scale, because no scale is sensitive enough. Indeed, even an ounce of any element is known to contain billions and billions of atoms. What Dalton did do, however, was to give a weight to one atom in terms of the weight of another. If atom A is arbitrarily assigned a weight of 1 atomic weight unit, then atom B might be assumed to weigh twice as much, and therefore be assigned an atomic weight of 2 units. Another atom, C, if it were assumed to be four times as heavy as atom A, would have an atomic weight of 4, and so on.

The remarkable thing about all this, and the thing that justified it, was that the atomic weight idea could be used to explain the law of definite composition. Consider the white and green sphere molecules you have already made. Let us assume each green atom to be four times as heavy as each white one. If we assigned each white atom an atomic weight of 1, this would mean an atomic weight of 4 for each green one. Now, proceed to make six green–white molecules. Notice that each such molecule contains a green and a white atom, and that it also has a definite weight make-up. Because of our atomic weight assignments, each molecule could be said to weigh 5 atomic weight units. One of these would be due to the white atom, of atomic weight 1, and the 4 others to the green atom, of atomic weight 4. Recall that the toothpick or dowel-rod bond must be considered weightless.

The entire sample of six molecules would contain 6 atomic weight units of white (6×1), and 24 units of green (6×4). Notice that the ratio of white to green remains fixed at 1 to 4. This is because $\frac{1}{4} = \frac{6}{24}$. Furthermore, if we split the six molecules into two samples, of, say, two and four, this ratio still holds, and is the same for each group. Here, the first group would be 2 parts by weight white, and 8 parts by weight green. The second group would be 4 parts white and 16 parts green, but $\frac{2}{8} = \frac{4}{16}$, because they each equal $\frac{1}{4}$.

What you have done, of course, is to show that any sample of this green–white compound must have the same composition by atoms, and therefore by weight, as any other sample. In other words, you have explained the law of definite composition.

PREDICTING NEW LAWS

So far, we have used the atomic theory to help us understand the laws which we already know exist. The beauty of the atomic theory—indeed of any successful theory—is that it also enables scientists to predict new laws which have not yet been discovered. Let us see how, by means of his theory, Dalton was able to predict a third weight law; one that was subsequently found to be true.

Begin by selecting three white and two red spheres, as well as three toothpicks, or dowel rods. Using charcoal, write the letter "A" and the number "1" on each of the three white spheres. This will identify them as atoms of element A, each having an atomic weight of 1. Next use white chalk to mark each red sphere with the letter "B" and the number "6". This indicates that they are atoms of a second element, B, each of atomic weight 6. Each B atom, in other words, is six times as heavy as each A atom.

Now, using one white and one red sphere, connect them to opposite ends of a toothpick, or dowel to make what we may call the two-atom molecule AB. Chemists call this the **formula** of compound AB, and it tells them the atomic makeup of its molecules. This molecule is similar to those we have made before, and it certainly does represent one molecule of the simplest possible compound of the elements A and B. Notice that the two laws we have just spoken of hold here. The AB molecule does indeed weigh 7 atomic weight units, which is the sum of the atomic weights of atoms A and B. This assumes, of course, a weightless dowel-rod bond. Notice also that molecule AB is 1 part by weight A and 6 part B. One billion such molecules would contain 1 billion parts by weight of A and 6 billion parts of B. Hence conservation of matter and definite composition hold for this compound, as indeed we know they must, and this is fine.

Next, suppose we find that a second compound of elements A and B exists, with different properties from those of the simple one we have been considering. This is actually often the case. The elements carbon and oxygen, for example, are known to form two different gaseous compounds: carbon monoxide, a deadly poison; and the harmless gas carbon dioxide, used to extinguish fires. The elements hydrogen and oxygen form the well-known compound, water; but they can also form the compound hydrogen peroxide, used as a bleaching agent and disinfectant.

To see how our sphere models might account for this, make a second white–red AB molecule exactly like the first. Notice that a second hole is still available in the red wooden sphere. If using styrofoam, simply make a second hole yourself. Into this, place a second toothpick or dowel rod, and attach a second white sphere into this rod's free end. You have now made a second molecule, containing only A and B atoms, but this one might better be identified as an A_2B molecule. Its formula indicates that it contains two A atoms and one B atom, rather than one of each. True, this is more complicated than the first, but it is just as reasonable. Furthermore, it explains how two different compounds of elements A and B might both exist. Since the two molecules are different, it follows that compounds made up of billions of each would also be different.

Finally, some careful thought should convince you that something else must also follow—something connected with the weight make-up of each compound. We have already noted that the AB molecule is 1 part A and 6 parts B, by weight. The A_2B molecule must therefore be 2 parts A and 6 parts B, because it contains just twice as many A atoms, namely 2. Consequently, one billion such A_2B molecules must be 2 billion parts by weight A and 6 billion parts B. This is a direct outcome of the assigning of atomic weights, and the building of the two different molecules. In other words, compared to fixed weights of B (in this case 6), the weights of A which will combine with these fixed weights of B must be in a ratio of 1 to 2.

When such pairs of compounds are analyzed, they are indeed found to show similar relationships. Three grams of carbon will combine either with 4 grams of oxygen to form carbon monoxide, or with 8 grams of oxygen to form carbon dioxide. One gram of hydrogen is known to combine with 8 grams of oxygen to form water, but it may also combine with 16 grams of oxygen to form the compound hydrogen peroxide. In both of these cases, the ratio of weights of the second element which can combine with fixed weights of the first is 1 to 2. Now obviously the ratio need not always be 1 to 2, because the atomic make-up of the molecules might not always change from 1 to 2. The ratio should, however, always be a small whole number one, since the new molecule might contain 3 or 4 A atoms—or perhaps a different number of B atoms. Atoms are assumed unbreakable in all their chemical combinations, however, and hence no molecule could contain a fraction of an atom. This is why the ratios must be small and whole, and why the atomic theory was able to predict what became known as the **law of multiple proportions**.

SIMPLE MOLECULAR MODELS

Introduction

Again, as in Unit 2, the reader is advised to use an organic chemistry student ball-and-stick kit. The one previously described in Unit 2 Introduction is ideal.

Readers still preferring to use styrofoam spheres are referred to the instructions already given (Unit 2, Introduction) for their preparation. As indicated in the text (p. 4), a pipe-cleaner, cut in short lengths, can serve in place of toothpicks. The pipe-cleaner is especially useful when "bent" bonds are required.

The models described in this unit are designed to give the reader a "feel" for the technique of representing simple molecules by joined-together spheres. No attempt has been made, at this stage, to be accurate—either with respect to scale or bond angles. These problems will be faced in Units 5, 6 and 7.

Formulas and Valence

So far, we have not bothered to assign specific elements for our spheres to represent. Now it is time we did. As we have already indicated, each sphere you have used represents one atom of a particular element. Five differently colored spheres, therefore, represent atoms of five different elements. It is an arbitrary but convenient convention that black spheres represent carbon atoms. White spheres stand for hydrogen atoms. Red ones are oxygen atoms. Blue spheres represent nitrogen atoms, and light green ones represent chlorine atoms. Although we may occasionally

indicate a special role for a particular sphere, the above assignments are usually adhered to by chemists.

Some of these represented elements are no doubt already familiar to you. Hydrogen, the lightest element, is a dangerously
inflammable gas. Oxygen and nitrogen are the principal com-
ponents of our atmosphere. Chlorine is a greenish-yellow,
poisonous gas—the only one of these four gaseous elements that
has a color or an odor. Carbon is most familiar to us as the
charcoal for the picnic fire, or as the graphite that is misnamed the
"lead" of lead pencils. Diamond, too, is a form of the element
carbon, even though it does not resemble either charcoal or graphite
in physical appearance. This might appear to you to be contrary to
what has been said so far. Actually it is not, and you may wish
to look ahead to Unit 7 to see how more detailed models of these
substances help explain this.

From their studies, chemists have learned a great deal about
the various compounds these five represented elements can
form.

Among other things, they have determined the formulas
and geometric shapes of most of these compounds' molecules.
You can use this information to construct models of some of these
molecules, using the techniques already described in the previous
pages. Let us now turn to such constructions.

BUILDING MOLECULAR MODELS

Hydrogen chloride. This compound is a gas at ordinary temperatures and is extremely soluble (easily dissolved) in water. The
resulting solution is known as hydrochloric acid, a very dilute
solution of which is present in our stomachs. It has the formula
HCl, indicating that its molecules each contain one atom of
hydrogen (symbolized by the letter H) and one atom of chlorine
(symbolized by the letters Cl). It is therefore of the AB type, as
simple a molecule as is possible between two dissimilar atoms. To
model it, simply connect one white and one green sphere by
means of one short dowel rod or toothpick (Fig. 1).

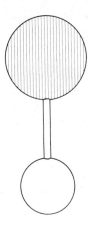

Fig. 1. Model of the hydrogen chloride molecule, an example of an AB type compound.

Molecules of elements. Both hydrogen and chlorine exist, in their elementary form, as two-atom units, rather than as single atoms. Join two white spheres with a short dowel, or toothpick, to represent a hydrogen molecule (formula H_2), and then join two green spheres to each other with a long dowel, or toothpick, to indicate one molecule of the element chlorine (formula Cl_2). Since these units both contain more than one atom (Fig. 2), they too are called molecules. They, however, are molecules of elements, because they each contain only one kind of atom.

Water, a compound of hydrogen (H) and oxygen (symbol O), has the familiar formula H_2O. This tells us that each of its molecules contains two hydrogen atoms and one oxygen, a total of three atoms, which makes it identical to our already constructed A_2B molecule. To model it, insert short dowel rods or toothpicks into each of the two holes of a red sphere (these holes should be about 105° apart) and join a white sphere to each of the dowels' free ends (Fig. 3).

Ammonia is a choking gas which is even more water soluble than hydrogen chloride. The solution made by dissolving it in

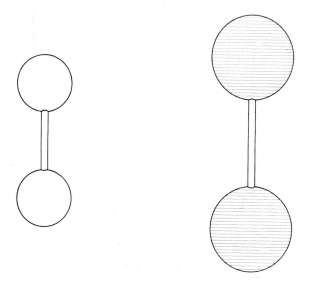

FIG. 2. Models of hydrogen (left) and chlorine (right) molecules. They are considered elements because they each contain only one kind of atom.

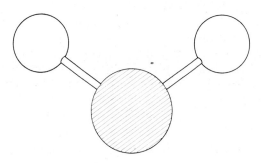

FIG. 3. Model of water molecule, H_2O, an example of an A_2B type compound.

water is, appropriatcly enough, known as ammonia water. The household ammonia used for cleaning is simply a dilute solution of ammonia water. The formula for ammonia, NH_3, tells us that each molecule contains one atom of nitrogen (symbol N) and three atoms of hydrogen (H), and hence it is more complicated than our previous models. To make it, insert three short dowels into the three holes found in a blue wooden sphere, and attach a white sphere to the free end of each dowel (Fig. 4). Or, if using styrofoam, make three holes at the angles shown in Fig. 4, and use toothpick connectors.

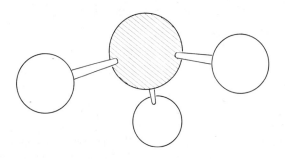

Fig. 4. Model of ammonia, NH_3.

Methane is best known as the principal component of the natural gas supplied to most kitchen stoves for cooking. Its molecules are even more complex than those of ammonia, as its formula, CH_4, immediately reveals. Here, one carbon atom (symbol C) is attached to four hydrogen (H) atoms. To construct the molecule, place short-length dowel rods into each of the four holes of a black sphere. Next attach white spheres to the free ends of each dowel (Fig. 5). If using styrofoam, use toothpick connectors, placed in the sphere at angles as shown in Fig. 5.

In the construction of all molecular models subsequently described in this manual, we shall follow the convention of joining a hydrogen "atom" to any other "atom" with a short-length dowel. All bonds between all other "atoms" should be made with the longer dowels. This and the color conventions are strictly adhered to in most of the molecular models made and used by chemists. If you adopt them now, you will be better able to recognize and understand any models you may encounter later on in your studies.

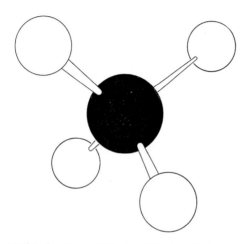

Fig. 5. Model of methane, CH_4, in which four H atoms are joined to a central carbon atom.

THE CONCEPT OF VALENCE

If you examine the four compound molecules you have just made, you will see that, in each case, one atom of an element is joined to different numbers of hydrogen atoms. It is as if these different atoms have the ability to combine with different numbers of hydrogen atoms. It is easy to see how the idea of combining power of the different elements' would naturally follow the determination of the formulas of their compounds.

The four molecules you have made are typical in the sense that the combining powers shown by the elements in them are the same as they show in most of their other compounds. Thus both hydrogen and chlorine have the ability to combine with one atom of hydrogen, as shown in the models of hydrogen chloride and molecular hydrogen. Oxygen, on the other hand, is able to combine with two hydrogen atoms, as the water molecule shows. Nitrogen can combine with three and carbon with four hydrogen atoms, as the ammonia and methane models reveal. Chemists have given the name **valence** to this combining power, and a simple definition of the valence of an element would be the number of hydrogen atoms with which one atom of that element can combine.

By this definition, the valences of hydrogen and chlorine would each be 1. Oxygen would have a valence of 2. Nitrogen has a valence of 3, although in some of its other compounds it exhibits other valences as well. Carbon, because it combines with four hydrogen atoms to form methane, is said to have a valence of 4. As you can see, the number of holes, and therefore the number of dowel rods insertable in a given "atom", also shows its valence. Table 1 lists the valences, as well as other pertinent information, about the five elements represented so far in our work.

TABLE 1. THE ELEMENTS REPRESENTED IN YOUR MODELS

Element	Symbol	Atomic weight	Valence	Brief description	Suggested color of sphere
Carbon	C	12	4	black solid	black
Chlorine	Cl	35.5	1	greenish, poison gas	green
Hydrogen	H	1	1	inflammable gas	white or yellow
Nitrogen	N	14	3	unreactive gas	blue
Oxygen	O	16	2	gas that supports burning	red

USING VALENCES TO MAKE NEW MOLECULES

Once you have developed the concept of valence as the number of hydrogen atoms with which a given atom can combine, and have assigned a valence of one to each hydrogen atom, the number of hydrogen atoms becomes equivalent to the number of bonds this given atom is capable of forming. This follows because each hydrogen atom requires one bond to hold it. Our model spheres show this nicely by the number of holes they contain, which, in turn, limits the number of dowel rod, or toothpick, bonds they can make with other spheres. If you make the additional assumption that each atom tries to combine until all its valence bonds are joined to other atoms, it then becomes possible to predict the formulas, and therefore construct models of all kinds of new compounds even before these formulas are checked by chemical analysis. This last assumption is reasonable because we do not find HO molecules, but only H_2O ones. This would seem to indicate that O atoms do not stop combining until they have filled both their valence bonds with H atoms.

Let us try to construct the model of a molecule of carbon and chlorine atoms, making use of the above ideas. We know the valence of carbon to be 4, as the four holes in each black sphere remind us. Insert 4 long dowels into one such sphere's holes, and finish by adding the proper number of green spheres to the dowels' free ends. Since each green sphere has one hole (a valence of 1), it is obvious that one goes on each dowel to give the molecule CCl_4, containing one carbon (C) and four chlorine (Cl) atoms (Fig. 6). Or, use styrofoam and toothpicks, as shown in Fig. 6. From the atomic weights given in Table 1, we see that the composition by weight of such a molecule must be 12 parts carbon (12×1) and 142 parts chlorine (35.5×4). If we then analyze the compound, known as carbon tetrachloride, which finds wide use as a cleaning agent, we find that it agrees with this predicted composition.

Now let us try modeling a molecule of a compound of carbon and oxygen. We might begin again by placing four dowels into

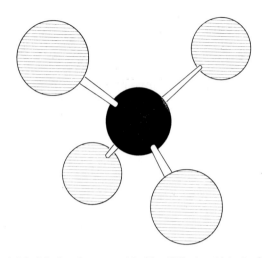

Fig. 6. Model of carbon tetrachloride, CCl_4, in which the four H atoms of methane have been replaced by four Cl atoms.

the four holes of a black sphere. Each oxygen atom, as represented by a red sphere, has a valence of 2, and therefore needs two dowels to fill its two holes. Since it is physically impossible to bend the dowels around to fit into these two holes, let us replace the dowels with four of the metal springs in your kit. Or, if using styrofoam, replace toothpicks with short lengths of pipe-cleaner. The springs, or pipe-cleaner, may be bent around, and a pair may be inserted in each of two red spheres. Here we have to stop because all valence bonds have been used up. The required molecule therefore has the formula CO_2, which indicates its makeup to be one carbon and two oxygen atoms (Fig. 7). Notice that in this molecule there are two bonds between each atom. These are known as double bonds to distinguish them from the one-dowel single bonds we have used so far. The compound whose molecule we have just modeled is, of course, carbon dioxide, the gas which puts out fires. When it is analyzed, it does indeed yield the expected result: 12 parts carbon (12×1) and 32 parts oxygen (16×2)

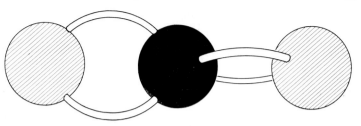

FIG. 7. Model of carbon dioxide, CO_2, showing double (bent) bonds.

by weight. Note that 12 to 32 may be reduced, by dividing each by 4, to 3 to 8, which is the analysis given earlier in the manual for this compound.

Hydrogen peroxide, the less common compound of hydrogen and oxygen, has the formula H_2O_2. If you experiment with the various ways of joining two white and two red spheres so as not to have any bonds left over, you will end with the molecule pictured in Fig. 8. Comparing this with the already constructed water

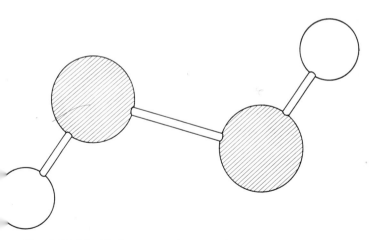

FIG. 8. Model of hydrogen peroxide, H_2O_2. Note that each oxygen atom still forms two bonds, each hydrogen atom one bond.

TABLE 2. STRUCTURAL FORMULAS OF COMMON ELEMENTS AND COMPOUNDS

Name of substance	Molecular formula	Structural formula
Hydrogen	H_2	H—H
Chlorine	Cl_2	Cl—Cl
Hydrogen chloride	HCl	H—Cl
Water	H_2O	H | H—O
Hydrogen peroxide	H_2O_2	H—O—O—H
Ammonia	NH_3	H | N—H | H
Methane	CH_4	H | H—C—H | H
Carbon tetrachloride	CCl_4	Cl | Cl—C—Cl | Cl
Carbon dioxide	CO_2	O=C=O

Single bonds are represented by a dash: —
Double bonds are represented by a double-dash: =

molecule (Fig. 3) reveals that the peroxide has two, rather than one, oxygen atoms. By referring to the atomic weights of Table 1, we see that while water has the composition by weight of 2 parts hydrogen (1×2) to 16 parts oxygen (16×1), hydrogen peroxide would be 2 parts hydrogen (again, 1×2) to 32 parts oxygen (16×2). If we simplify each of these ratios by dividing each number by 2, we get 1 to 8 water, and 1 to 16 for the peroxide. You may recall these ratios in our experiment on the prediction of new laws, for they are an example of the law of multiple proportions. Here, a fixed weight of hydrogen (2 parts) may combine either with 16 parts or 32 parts of oxygen. Sure enough, the 16 and 32 are in the ratio of small whole numbers: namely, 1 to 2.

STRUCTURAL FORMULAS

Now that you have made molecular models of several typical compounds, you are no doubt aware of the advantages of the actual models over molecular formulas. For one thing, the models tell you how the atoms are joined to each other, the formulas do not. The formula H_2O_2, for example, does not show whether the H atoms are joined to each other, to one O atom, or to both. The model does, and is therefore preferable, providing one has the time and materials to build it. Often, however, a chemist has neither commodity, and so he compromises by drawing the three-dimensional model on two-dimensional paper, without ever building it. Thus, H_2O_2 could be written as H—O—O—H. Of course, this does not show the proper shape of the molecule because we have lost a dimension in transferring to the paper, and this is the great limitation of such so-called structural formulas. It does show how the atoms are connected, without going through the fuss and bother of building the actual model. How many of the already constructed molecules' structural formulas can you draw? You might try, and then compare your answers with those given in Table 2.

WAVE MECHANICAL MODEL OF ATOMS

Introduction

The models whose uses are described in this unit can be made easily and at low cost from styrofoam spheres and teardrop-shaped styrofoam, plus toothpicks and/or thin dowel rod (drug-store-type medical applicators, sold as thin pieces of dowel whose ends can be wrapped in cotton, are ideal) and water-soluble glue.

These models represent the so-called orbitals of quantum mechanics, of which we shall only consider two: the *s*- and *p*-type of orbital. Quantum mechanics tells us *s*-orbitals are spherical, and hence are nicely represented by the styrofoam spheres themselves. *P*-orbitals can be represented by the tear-drop-shaped styrofoam. A single *p*-orbital (Fig. 9) is modeled by joining two such teardrops at their thin ends, using a toothpick as connector.

An approximation to the three mutually perpendicular *p*-orbitals chemists talk of may be modeled as follows. Draw the pattern shown in Fig. 10 on a piece of light cardboard, using compass and protractor. Draw all lines before cutting out the circle with a razor. The diameter of this circle should be the same as that of an available styrofoam sphere: $1\frac{1}{2}$ in. is convenient. Push the sphere into the cut-out cardboard circle, as shown in Fig. 10, and make pencil marks in the sphere where the four "mark" lines of the cardboard touch it. Then gently rotate the sphere 90° about its *c–d* axis, as it sits in the cardboard, until the pencil marks originally made at "*a*" and "*b*" are midway

(a) p-orbital

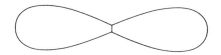

(b) Assembling the model

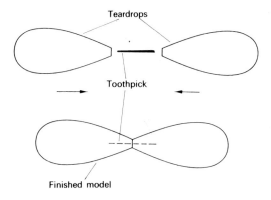

Teardrops

Toothpick

Finished model

Fig. 9. (a) *p*-orbital; (b) assembly of *p*-orbital model from teardrops.

between these points. When they are, make additional pencil marks where "*a*" and "*b*" marks now touch the sphere, giving six marks in all.

If one imagines the sphere's center to be at the intersection of the *x*-, *y*-, and *z*-axes shown in Fig. 10a, the marks on the sphere should correspond to where these axes would emerge from the sphere. Using a toothpick as a drill, carefully punch holes in the sphere, at each marked spot, directing them toward the sphere's center. To check accuracy in this last operation, push toothpick slightly into sphere, and grasp it as shown in Fig. 11c, rotating toothpick as shown about a horizontal axis of rotation. Look at sphere during rotation—if sphere "wobbles" from side to side,

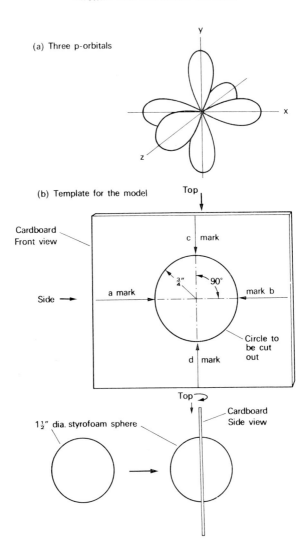

(a) Three p-orbitals

(b) Template for the model

FIG. 10. (a) 3 *p*-orbitals; (b) cardboard jig for the three *p*-orbitals model.

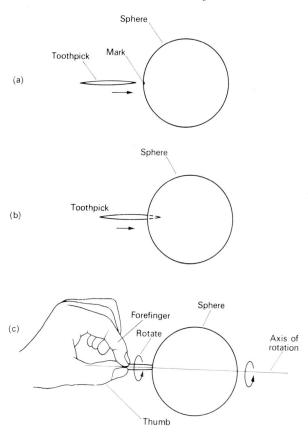

FIG. 11. (a), (b) and (c) Checking for accurate centering of sphere holes.

toothpick is not headed toward center, and must be removed and reinserted until no sphere "wobble" is noticeable during such rotation.

When all six marks have been "drilled" in the sphere, as described above, cut a bit off the thin ends of each of six tear-drops, inserting a toothpick halfway into each cut off end (Fig. 12). A dab of water-soluble (Elmer's) glue, inserted on end of toothpick before insertion, will help hold it firmly in place. Then push the toothpick ends of these teardrops into the pre-prepared holes in your sphere (Fig. 12), again using glue for a firmer model. What you have modeled represents one *s*-orbital (the sphere) surrounded by three *p*-orbitals (the teardrops), or at least a reasonable approximation to it.

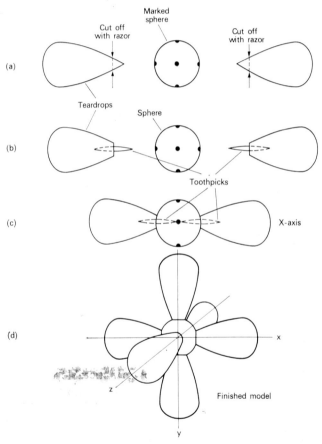

FIG. 12. (a)–(d) Constructing model of an *s*- and three *p*-orbitals.

While the model can stand unpainted, you may wish to paint the three *p*-orbitals different colors, to distinguish the one along the *x*-axis from the one along the *y*-axis, and so on. Remember, in this case, to use water-soluble, latex base paint. The finished model should resemble Fig. 12d. Remember it is an approximation, designed to show approximate shapes and orientations of the orbitals, and is not necessarily to scale.

This unit, and those that follow, will appeal more to readers with some background in chemistry. No attempt at development of this background has been made here, but interested laymen are referred to the bibliography given in the Appendix.

The Meaning of Orbitals

From quantum mechanics comes the concept of "orbitals", where an orbital is thought of as that portion of space in which a given electron of a particular atom is likely to be found. Since these orbitals extend, at least theoretically, infinitely far from their atom's nucleus, chemists usually compromise by imagining an orbital of finite size to represent the electron's "residence", say 95% of the time. Thus any model of an orbital, since it must be of finite size, is obviously not able to represent more than, say, 95% of it.

The models we shall consider here also cannot indicate anything about the probability of finding an electron at a particular spot within an orbital, for they only represent the boundary, or—more precisely—95% of the boundary, of the orbital. The models do, however, illustrate the approximate shapes and directions of the *s*- and *p*-orbitals, and it is this information which we shall attempt to make use of here.

THE SHAPES OF MOLECULES

If, as chemists believe, chemical bonds are made when two electrons, originally from two separate orbitals, merge to share both orbitals, it seems reasonable to assume that the shapes of

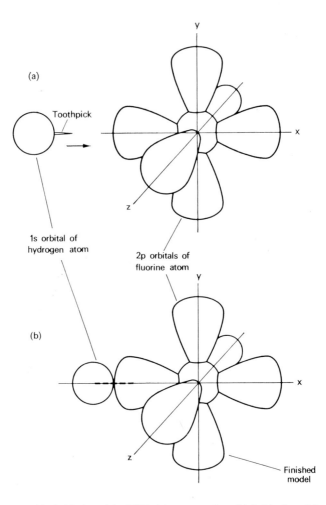

F𝘪𝘨. 13. Orbital model of HF: (a) construction, (b) finished model.

these orbitals must influence the shape, and hence the bond angles, of the resulting combinations. It is instructive, in this connection, to imagine the formation of several such simple combinations, and to then compare the thus predicted bond angles with experimentally determined values.

As we do this, keep in mind that a given atom is imagined to consist of a central nucleus, surrounded by electrons. Our orbital models represent the orbitals of such an atom's least tightly held electrons, while its nucleus and securely held electrons must be imagined buried inside the model, perhaps at its center.

ORBITAL MODEL OF HYDROGEN FLUORIDE, HF

Hydrogen, with one electron in its so-called 1s orbital, can be modeled by a styrofoam sphere ($1\frac{1}{2}$ in. diameter is convenient). Fluorine has both 1- and 2s orbitals, as well as three 2p orbitals. The model shown in Fig. 12d represents fluorine: the p-orbitals by the teardrops, and the s-orbitals by the central sphere. Its surface would represent that (95%) boundary of the 2s, and the 1s orbital would then be a smaller sphere and buried inside, but sharing the same center as, the sphere in the model.

To model the hydrogen fluoride molecule, simply join a sphere to the orbital model of Fig. 12d, using a toothpick connector as shown in Fig. 13. Since the center of the "hydrogen atom" and the center of the "fluorine atom" both lie on the same straight line (the x-axis of Fig. 13), a linear molecule is predicted. This is in agreement with experiment.

ORBITAL MODEL OF WATER, H_2O

This model is identical to the one for hydrogen fluoride, except that a second "hydrogen" is attached to a teardrop along the "y"-axis, as shown in Fig. 14. Now it is seen that the bonds between the two "hydrogen atoms" do not lie on a straight line, but are bent, at an angle of approximately 90°. Thus water should be a bent molecule, as confirmed by experiment. While the

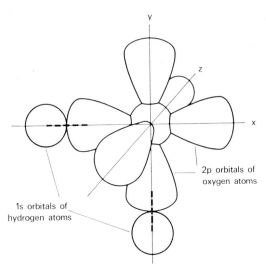

FIG. 14. Orbital model of H₂O.

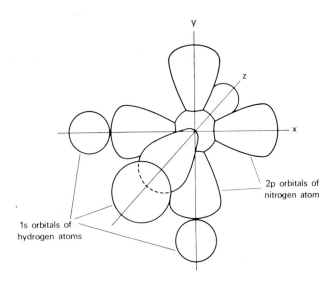

FIG. 15. Orbital model of NH₃.

bond angle in water is 104°, its analogs, H_2S, H_2Se and H_2Te, do approach the predicted 90° angle.

ORBITAL MODEL OF AMMONIA, NH_3

This model is made exactly as was the water model, except that a third sphere, representing a third "hydrogen atom", is attached to either one of the "z"-axis teardrops, to give the model shown in Fig. 15. Here, the three "hydrogen atoms" are each perpendicular in their bond angles to the other two. Thus the expected shape of the ammonia molecule is pyramidal, the three "hydrogens" forming the base, and the "nitrogen atom" the apex, of a pyramid.

The experimentally determined shape of ammonia is, indeed, pyramidal, but with the bond angles between hydrogens about 104° instead of the expected 90°. Again, however, the analogs of ammonia, PH_3 and AsH_3, have bond angles much closer to the expected 90° than this.

USE OF ORBITAL MODELS

Naturally the models discussed above are most useful to chemists, and especially to students of chemistry trying to visualize the influence of bonding orbitals on molecular shape. Each of these models represents, of course, all the analogs of the compound being considered, since the same bonding orbitals are involved. Thus, HF is similar in orbital shape to HCl, HBr, HI; H_2O resembles H_2S, H_2Se, H_2Te; and NH_3 would be like PH_3 or AsH_3.

While the modeling and use of d-orbitals is not considered here, the techniques are similar. Interested readers are referred to the Appendix bibliography.

UNIT 5

ACCURATE MOLECULAR MODELS

Introduction

All models described in this unit are best made from styrofoam spheres. They can be "bonded" together with toothpicks whose ends have been dipped in water-soluble glue, or, for the so-called lattice models we shall describe, by short lengths of dowel rod such as come in drug-store medical applicators. If sphere is to be joined directly to sphere, the surfaces to be joined can be held with water-soluble ("Elmer's") glue, and are better left unpainted.

All exposed surfaces on the models can be painted, but only with at least a first coat of water-soluble, latex-base paint, brushed on so as to fill any pores. Tempera colors, when dissolved in white shellac (in place of water) as the vehicle, also have proven satisfactory. A second coat of enamel can often cover the latex-painted spheres, for the latex serves to protect the styrofoam from the dissolving action of the enamel. Readers who do plan to enamel their models would do well to proceed cautiously by experimenting first with a spare sphere before painting the model itself.

What color to paint each sphere is, of course, arbitrary. The code suggested (page 41) is widely used by chemists, but it is by no means rigid.

All other details of construction of the models will be discussed in the text of this unit. As in Unit 4, no attempt will be made to present the chemistry background assumed in the

40

discussion. Interested lay readers are referred to the bibliography given in the Appendix.

Element represented	Suggested color
All metals except copper, gold	Silver
copper, gold	Gold
Hydrogen	White or yellow
Carbon, silicon	Black
Nitrogen, phosphorus	Blue*
Oxygen, sulfur	Red*
Halogens	Green*

* Heavier members of a family can be made a lighter shade of the color being used.

BALL-AND-STICK (LATTICE) MODELS

The ball-and-stick models already discussed in Unit 3 are examples of so-called "lattice" models, except that the term lattice usually refers to an extended portion of a crystal, rather than to an isolated molecule. A "lattice" model of a single molecule is usually referred to as a ball-and-stick model. Regardless of what one calls it, this type of model is most useful in showing the angles between chemical bonds, and the location of the nucleus, or center, of each atom within the molecule.

The ball, or sphere, in such a model, therefore represents only the location of a particular atom, but not the entire volume of the represented atom. In the volume sense such models are not made to scale, although the distances between centers of the spheres, and the angles between dowel rod "bonds", can be made fairly accurate. The dowel connectors, of course, represent the chemical bonds between the atoms.

SPACE-FILLING (PACKING) MODELS

Although it is not easy to fix the sizes of atoms, chemists do talk of them as if they had fixed boundaries. What they mean, of course, is that even the least tightly held of an atom's electrons have, say, a 95% probability of being within a particular segment of space. We met this idea in our consideration of atomic orbital models in Unit 4. While these orbital models (i.e. Figs. 13–15) may accurately represent electron distribution within a molecule, it is far easier, and apparently accurate enough to represent atoms as hard spheres, regardless of which orbitals their least tightly held electrons may occupy.

The surface of a styrofoam sphere therefore represents the boundary of a given atom. The atom's nucleus is imagined to be at the geometric center of the sphere. The size of the sphere, its radius, will depend upon the scale you choose, and on the closest approach of atoms in the solid state. The actually measured quantities, using solid iodine as an example, are internuclear distances. In the iodine crystal, each iodine atom's nucleus is 2.67×10^{-8} cm away from one other atom's nucleus, and 4.30×10^{-8} cm away from its next closest neighbor's nucleus. This latter value is taken as the distance of closest approach between non-bonded atoms, so that half this distance, 2.15×10^{-8} cm, is taken as the effective radius, called the van der Waals radius, of an iodine atom (Fig. 16).

The shorter internuclear distance observed in the above iodine example is assumed to be the distance between bonded-together atoms of an I_2 molecule (see Fig. 16), so that half this is taken as the so-called covalent radius. The necessary interpenetration of spheres representing the atoms within a molecule is then the model's way of showing that each electron making the chemical bond belongs in part to its nearest neighbor atom, as well as to its own atom.

The scale you choose for your models, which then determines the degree of magnification, is of course arbitrary. Letting 1 in. = 1 Å (1×10^{-8} cm) makes models a bit large, while allowing 1 cm = 1 Å results in models being too small to be easily seen or worked on. A suggested compromise would be to let 2 cm = 1 Å, resulting in a linear magnification of 200 million over the imagined actual size of the atoms. This is the scale we

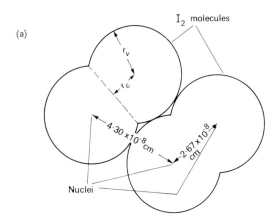

(a)

Van der Waals radius, $r_v = 2 \cdot 15 \times 10^{-8}$ cm

Covalent radius, $r_c = 1 \cdot 33 \times 10^{-8}$ cm

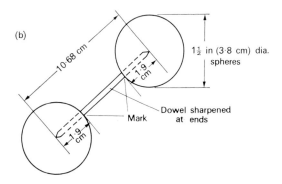

(b)

FIG. 16. I_2 models: (a) space-filling with R_v and R_c; (b) ball-and-stick.

shall adhere to in this manual, but readers should feel free to depart from it to suit their convenience.

The usual procedure is first to look up such things as bond angles, and van der Waals and covalent radii, for atoms of the molecule being considered. Suitable references are given in the appendix. For ball-and-stick type models, a cardboard guide, or template (similar to the one shown in Fig. 10b), is made to insure that the dowel rod bonds are connected at the proper angles. The internuclear distance, when converted into the scale of the model, then determines the length of the dowel bond.

Space-filling models require an additional step to allow "inter-penetration" of bonded "atoms" within the molecule. In practice, this amounts to slicing off a portion of the sphere with a sharp, serrated, saw-like blade, and then making this cut surface smooth-ly flat by sanding it with coarse garnet paper. The size of these cuts is also predetermined by the van der Waals and covalent radii, and these flat portions of two "bonded-together" spheres can then be glued together with Elmer's, or some similar water-soluble glue. Some typical models, illustrative of these techniques, will now be considered.

MODELS OF THE IODINE MOLECULE, I_2

Having already used iodine to illustrate the various atomic radii, let us use it as our first example. The ball-and-stick model of the I_2 molecule (see Fig. 2) is identical in appearance to either the H_2, or Cl_2 models previously discussed in Units 2–3. The single bond eliminates the bond angle problem, so we need only determine the length of the dowel rod "bond". By our scale of 2 cm $= 1$ Å, this would be twice the internuclear distance of 2.67 or 5.34 cm. For ball-and-stick models, however, I suggest this be doubled, to 10.68 cm. The double scale here will soon be justified. If we choose $1\frac{1}{2}$ in. diameter spheres, as shown in Fig.

16b, we would cut a dowel 10.68 cm in length, and mark off 1.9 cm (the sphere radius) from each end. If the dowel ends are sharpened to a point with a razor, they will easily push into the spheres. Using the test described in Fig. 11, push each end of dowel half-way into each sphere, stopping at the 1.9 cm marks, to produce the finished model shown in Fig. 16b.

To make a space-filling model of iodine, the properly scaled spheres must each be sliced and flattened so as to fit together as shown in Fig. 16a. This is done to scale first on a piece of paper, and then transferred directly to the sphere. First, using a compass, draw a circle representing the van der Waals radius, 2.15 Å. On our scale, this would be 4.30 cm (Fig. 17a). Draw diameter AOB, and mark on it the covalent radius, OC. Given $r_c = 1.33$ Å, OC must be 2.66 cm (Fig. 17b). Draw DE perpendicular to AB at C (Fig. 17c), and measure the distance DB, opening a compass to this distance (Fig. 17d). With the compass open to this distance, draw a circle of diameter DB on each of two white, 6.8 cm diameter styrofoam spheres (Fig. 17e), and slice off these circles with a sharp, serrated knife (Fig. 17f). Finally, join the two spheres at their flat, sliced-off portions, with water-soluble ("Elmer's") type glue, to give one unit of the model shown in Fig. 16a. A rubber band will hold the two spheres together while the glue dries. Finally, paint the two sphere unit, as suggested in the introduction to this unit.

Except for slight variations in color and size, this model could represent any of the following molecules: F_2, Cl_2, Br_2, H_2, O_2. Their models would be constructed exactly as described above, except that distances OA and OC would correspond to the proper van der Waals, and covalent, radii. These values are given in the appendix.

MODEL OF THE HYDROGEN CHLORIDE MOLECULE, HCl

The construction of a space-filling model follows the general procedure described above. According to our scale (2 cm $= 1 \times 10^{-8}$ cm), and the accepted van der Waals radii (1.80 $\times 10^{-8}$ cm for Cl, 1.20 $\times 10^{-8}$ cm for H), we would select a 3.6-cm sphere

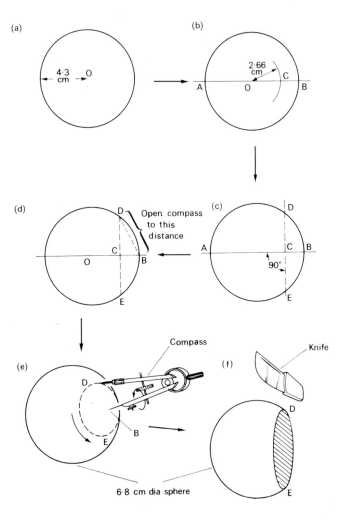

FIG. 17. (a)–(f) Construction of space-filling model of I_2.

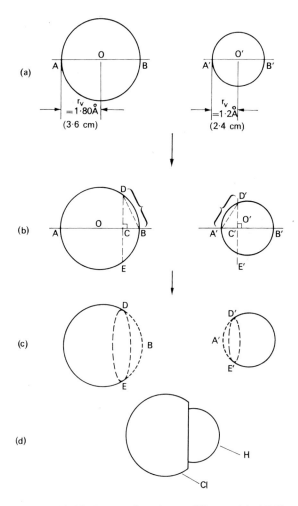

Fig. 18. (a)–(d) Construction of space-filling model of HCl.

to represent Cl, and a 2.4-cm one for the H atom. Now make two-dimensional drawings (i.e. circles) corresponding to the sizes of these spheres (Fig. 18a), and mark off on each's diameter *AB* its covalent radius, 1.98 cm on large, 0.60 cm on smaller sphere (*OC* and *O'C'* in Fig. 18b). Draw perpendiculars *DE* at *C* and *D'E'* at *C'* (Fig. 18b) and set your compass to the distance *DB*. Draw a circle of this radius on the large sphere, and repeat the procedure on the smaller sphere, using *A'D'* as the radius. Slice off these circled portions from each sphere, using a sharp, serrated knife, and making the cuts as flat as possible, finishing them with coarse garnet paper (Fig. 18c). Finally, join the spheres, using Elmer's, or an equivalent water-soluble glue, at their flat portions, to obtain the model shown in Fig. 18d.

The accurate ball-and-stick model of HCl (already illustrated in Fig. 1) illustrates the necessity of adopting a scale twice as large as that for space-filling models. This necessity arises because, if we wish to use spheres large enough to work with (1 in. or more), no dowel would show between spheres if we used the space-filling model scale of $2 \text{ cm} = 1 \times 10^{-8}$ cm. Hence the scale for our ball-and-stick model becomes $4 \text{ cm} = 1 \times 10^{-8}$ cm, so the 1.29×10^{-8} cm internuclear distance becomes 5.16 cm, the length of our dowel rod. Selecting $1\frac{1}{2}$ in. (1.9 cm) sphere for the Cl, and 1 in. (1.3 cm) for the H, we mark off these distances

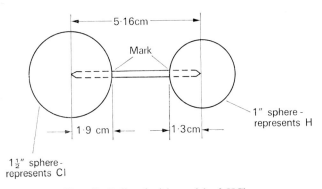

FIG. 19. Ball-and-stick model of HCl.

from the ends of the dowel (Fig. 19), sharpen each end, and insert them (using the wobble test, Fig. 11, p. 33) into our spheres up to the marks, producing the finished model shown in Fig. 19.

MODEL OF THE WATER MOLECULE, H_2O

Here, for the first time, we deal with a model whose central "atom" (the O) has more than one bond; hence we need to know the angle between these bonds, the so-called bond angle. In water, this is given as 105°. To represent oxygen, with its 1.40 × 10^{-8} cm van der Waals radius, we use a 5.6 cm diameter (2.8-cm radius) sphere. First, draw a circle of this diameter, and measure off an angle of 105° from its diameter AOB (Fig. 20a). The intersection of this angle with the circle, at B and B' (Fig. 20a) gives the distance, b, between points on the sphere representing oxygen, which are sliced off to accommodate the "hydrogens". Mark any two points on the large sphere, distance b apart (2.25 cm by our scale). Then, in the usual way, swing an arc from O, equal to the covalent radius of oxygen (1.32 cm on our scale) (Fig. 20b). Draw DE perpendicular to OB at C, and $D'E'$ perpendicular to OB' at C' (Fig. 20c). Now draw circles of radius DB at points B and B' on the large sphere, and slice these portions off in the usual way (Fig. 20d). Prepare two 2.4 cm radius spheres, representing the two hydrogen atoms, exactly as shown in Fig. 18 for the HCl molecule, and glue them into place, ending with the model shown in Fig. 20f.

The ball-and-stick model of water, and other multiple bond molecules, can best be made with the aid of a cardboard jig for accurate angle punching of the central spheres. Two such jigs are shown in Fig. 21. They can be made from small pieces of packing-case cardboard, by first marking the circles and angle guidelines indicated in the figure, and then cutting out the circles with a razor. Punches can be made from stiff wire ($\frac{1}{16}$-in. welding rod is ideal) by bending one end of the wire with pliers, as shown in the middle of Fig. 21, to form a handle. Snip the other end off so that, when pushed into the cardboard up to the

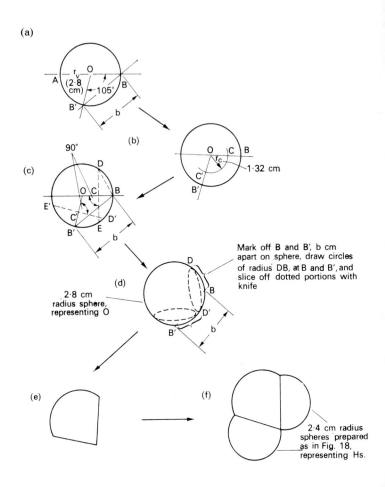

FIG. 20. (a)–(f) Construction of space-filling model of H_2O. O: $r_v = 1.40 \times 10^{-8}$ cm, $r_c = 0.66 \times 10^{-8}$ cm, bond angle 105°. H: $r_v = 1.2 \times 10^{-8}$ cm, $r_c = 0.30 \times 10^{-8}$ cm; two H's, each cut as in Fig. 18.

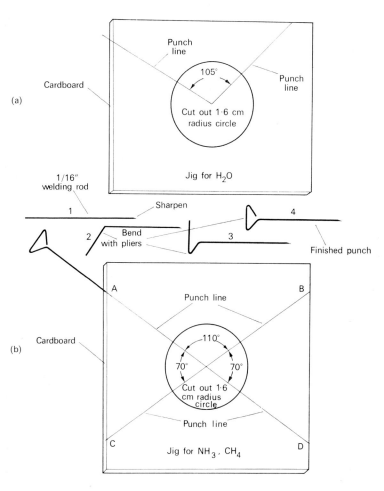

FIG. 21. Jigs for ball-and-stick models of (a) H_2O; (b) NH_3 and CH_4.

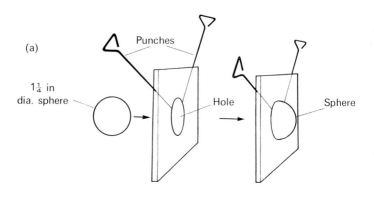

(a)

Punches

$1\frac{1}{4}$ in dia. sphere

Hole

Sphere

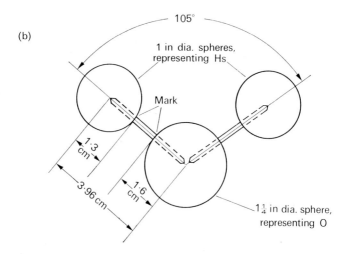

(b)

105°

1 in dia. spheres, representing Hs

Mark

1·3 cm

3·96 cm

1·6 cm

$1\frac{1}{4}$ in dia. sphere, representing O

FIG. 22 Ball-and-stick H_2O: (a) use of jig for O; (b) finished model.

handle, the other end is just at the center of the now cut-out circle, and sharpen this snipped off end with a file or knife sharpener. The punches are best forced in between the cardboard, using the punch lines to push them in accurately. Thus the thick cardboard actually holds the punches in place, as they are pushed in, and pulled out of, spheres that are inserted in the cut-out circles.

To model water, place a $1\frac{1}{4}$-in. diameter sphere in the H_2O jig so it is just centered. Some spheres have mold marks at their "equators" to guide you, or simply push sphere in until an equal volume sticks out of either side of the jig (Fig. 22a). Push the two punches in, then out of sphere, remove, and, being careful to center them (use the wobble test), push two marked and sharpened 3.96-cm dowel rods into each of two 1-in. diameter spheres. Finally, insert free ends of dowels up to the mark in the previously punched holes of the larger, $1\frac{1}{4}$-in. diameter sphere, to produce the finished model shown in Fig. 22b.

SPACE-FILLING MODELS OF AMMONIA, NH_3, AND METHANE, CH_4

Except for sizes, and an additional bond and H atom in methane, these two molecules can be considered identical, at least in construction of their models. To make ammonia, draw a 3.0-cm radius circle (the radius of the N-representing sphere), and draw angle BOB' at $109\frac{1}{2}°$ as shown in Fig. 23a. The distance BB', represented by b, will be the separation between centers of sliced-off circles on the large sphere. Mark off any two points, B and B', distance b apart, on your sphere (see Fig. 23d), and swing arcs of radius b from B and from B' to locate the third point, B'', where these arcs intersect. In making methane, the same procedure is used on the 1.54-cm radius carbon-representing sphere, but the arc-swinging from B and B' is repeated on the other (hidden in the Fig. 23d diagram) side of the sphere to locate the fourth needed point. In each case, circles of covalent radius DB (Fig. 23c), are arrived at in the usual way by swinging an arc from O (Fig. 23b) the length of this covalent radius, and

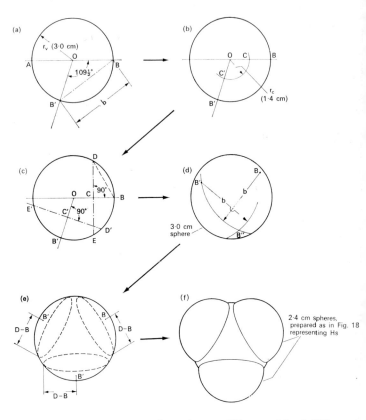

FIG. 23. (a)–(e) Construction of space-filling model of NH_3.
N: $r_v = 1.5 \times 10^{-8}$ cm, $r_c = 0.70 \times 10^{-8}$ cm, bond angles each
$109\frac{1}{2}°$. H: $r_v = 1.2 \times 10^{-8}$ cm, $r_c = 0.30 \times 10^{-8}$ cm; three H's,
each cut as in Fig. 18. For C: $r_v = 1.7 \times 10^{-8}$ cm, $r_c = 0.77 \times 10^{-8}$
cm. (f) Finished model.

then drawing perpendiculars at C and C' as shown in Fig. 23c.
Circles of this radius (DB) are then drawn on each sphere, using
each previously located point as a center, and these circles are
then sliced off with a sharp knife (Fig. 23e), leaving three flat
portions for N, and four for C. Finally, 2.4-cm spheres, repre-

senting H atoms, and prepared as in Fig. 18, are glued onto each flat portion, giving the finished models NH_3 (Fig. 23f), and CH_4.

BALL-AND-STICK MODELS OF AMMONIA, NH_3, AND METHANE, CH_4

These, too, are almost identical, and are made almost the same way. The central sphere for each (representing N and C) is best punched with the aid of the jig shown in Fig. 21b. In either case, a $1\frac{1}{4}$-in. diameter sphere is centered in the jig (Fig. 24a), and punched with the upper two (*A* and *B*) wire punches as shown in Fig. 24a. These punches are then removed, and the sphere rotated IN PLACE just 90° about its vertical y–y' axis (Fig. 24b), until the holes made by punches *A* and *B* are midway between the *A* and *B* punch marks, when viewed from above the jig. Then the sphere is punched at *C* only for the nitrogen, and at *C* and *D* for the carbon (Fig. 24c), so that the proper length dowel rods, suitably marked, sharpened, and pushed into 1-in. diameter hydrogen-representing spheres, can have their free ends inserted into these jig-punched holes. The resulting ammonia model is shown in Fig. 24d. Methane would be identical, except for a fourth "hydrogen" directly behind the upper one as shown in the figure.

The reason the jig shown in Fig. 21b can be used to prepare the above models is revealed from consideration of the geometry involved. In both the NH_3 and CH_4 molecules, the central atom is at the center, and the hydrogens at three, and all four, respectively, vertices, of a regular tetrahedron (Fig. 25). This solid can be inscribed in a cube (Fig. 26a) so that the geometric centers (*O* in the diagram) of both figures coincide. The dotted lines from *A*, *B*, *C* and *D*, to *O* (Fig. 26a) show the directions spheres representing N or C atoms must be punched, if one imagines the spheres centered at *O*. Note that "punches" *C* and *D* are at right angles to *A* and *B*, explaining the necessity of rotating the sphere 90° between punches (Fig. 24b). Figure 26b suggests an easy way of calculating the angle between punch directions, the so-called tetrahedral angle, AOB, of $109\frac{1}{2}°$. Solving for angle *X* of triangle

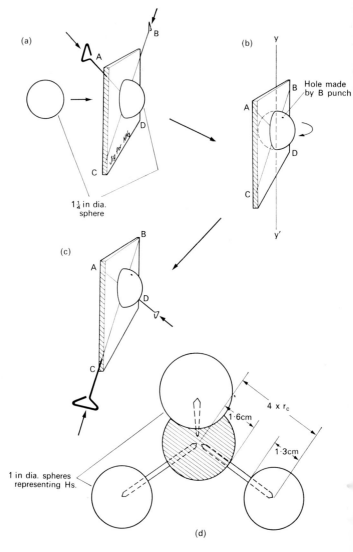

Fig. 24. (a)–(c) Use of jig for ball-and-stick NH₃ and CH₄;
(d) finished model.

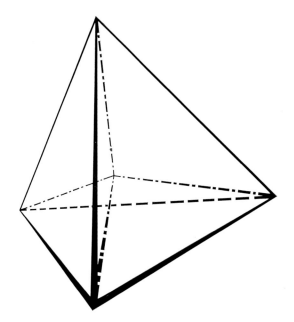

Fig. 25. Drawing of regular tetrahedron in which the geometric center is joined to the four vertices. Four congruent, equilateral triangles make up the faces.

OEB (Fig. 26b) is straightforward, and *X*, after all, is simply half of the angle we seek.

THE PHOSPHORUS MOLECULE, P_4

This four-unit molecule has P atoms at the four corners of a regular tetrahedron (Fig. 25), and nothing in the geometric center. This means that each P atom bonds to three others, and that the bond angles are 60°. For the space-filling model, select four 7.6-cm diameter spheres, and prepare each as follows. Draw a 3.8-cm diameter circle on paper, and measure off the 60° angle *AOE* (Fig. 27a). Swing an arc, r_c, representing the covalent radius,

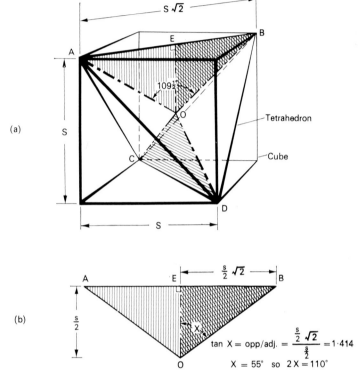

Fig. 26. (a) Tetrahedron inscribed in cube, showing geometry of
NH_3, CH_4 jig; (b) calculation of tetrahedral angle.

2.2 cm on our scale, from O, cutting diameter AOB at C (Fig
27b). Draw CD perpendicular to AB at C (Fig. 27b) and measure
the distance DB (represented by r in Fig. 27b) as the radius of
circle to be drawn on and then sliced off our sphere. The distanc
AE (represented by b in Fig. 27a) shows how far apart on the
sphere the centers of these circles must be.

On each sphere, therefore, mark off any two points, B and B'
b cm apart (Fig. 27c), and swing arcs b cm long from B and from
B'. Their intersection, B'', locates the third point. Now set you

compass to length *r* (Fig. 27b) and draw circles of this radius
directly on each sphere, with centers at *B*, *B'* and *B''*. Finally
slice off these marked circles (Fig. 27d), flattening the sliced-off
portions with garnet paper. Each sphere should look like Fig. 27e,
when viewed from directly below the cuts. Fit the spheres

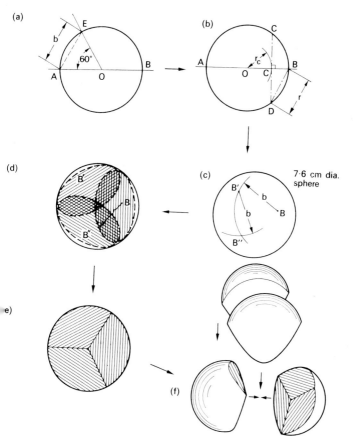

FIG. 27. (a)–(f) Construction of space-filling model of P_4. P: $r_v = 1.9 \times 10^{-8}$ cm, $r_c = 1.1 \times 10^{-8}$ cm (3.8 diam. = 7.6-cm diam. sphere).

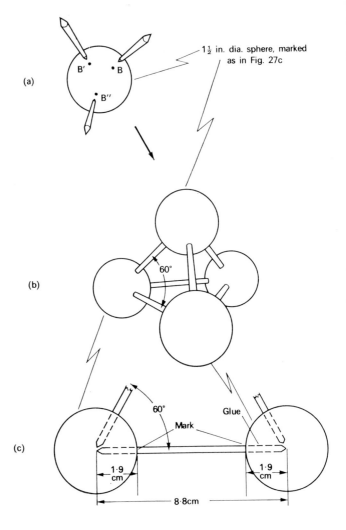

(a)

1½ in. dia. sphere, marked as in Fig. 27c

B′   B

B″

(b)

60°

(c)

60°

Glue

Mark

1·9 cm

1·9 cm

8·8cm

FIG. 28. (a) and (c) Construction of ball-and-stick P_4; (b) finished model.

together, first in twos as in Fig. 27f, and finally fit the two two-sphere units together as shown.

The ball-and-stick model of P_4 is probably better done without a jig. Use four $1\frac{1}{2}$-in. diameter spheres, locating points B, B' and B'' exactly as you did on the larger space-filling spheres (Fig. 27), but scaling your drawings down to the $1\frac{1}{2}$-in. diameter circle. Using our double-scale, cut six 8.8-cm lengths of dowel, marking 1.9 cm from each end, and then sharpening each end. Slowly insert three dowels into each sphere (Fig. 28a), using the wobble test described on p. 33 (Fig. 11), up to the 1.9-cm mark, and then withdraw them again. Finally, barely insert the dowels in the previously prepared holes, as shown in Fig. 28b, dab some glue on each end of dowel, between the sphere and the marking, and then slowly push dowels into the spheres up to the 1.9-cm markings.

BALL-AND-STICK MODEL OF SULFUR, S_8

The shape of this molecule (Fig. 29a) makes a space-filling model very difficult because of the additional slicing off of spheres, due to the compact shape of the eight-membered ring. The structure, however, can be better shown by means of a ball-and-stick model, as illustrated in Fig. 29a. To prepare it, select eight $\frac{1}{2}$-in. diameter spheres, and prepare each exactly as you did the $\frac{1}{4}$-in. diameter sphere shown in Fig. 22a. The jig here will differ 1) in that its cut-out circle will be of radius $\frac{3}{4}$ in., and (2) the angle between punches will be 100°, but will otherwise be identical to that shown in Fig. 21a. The two holes thus made in each sphere should be b (2.9) cm apart (Fig. 29c), a good check before proceeding any further.

Cut eight 8.3-cm lengths of dowel, mark 1.9 cm from each end, and then sharpen each end. Barely insert two dowels into each of our spheres (the bottom ones in Fig. 29a), and then barely insert their free ends into the other four (upper) spheres, to obtain a facsimile of Fig. 29a. When this is achieved, dab some glue on each end of each dowel, as shown in Fig. 29b, and then slowly push all dowels into spheres, up to the 1.9-cm marks.

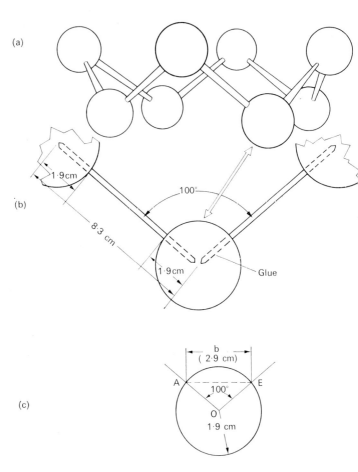

(a)

(b)

1·9 cm

100°

8·3 cm

1·9 cm

Glue

(c)

b
(2·9 cm)

A

E

100°

O

1·9 cm

FIG. 29. (a) Finished ball-and-stick S₈ model; (b)–(c) construction of model.

COMPLEX IONS

Some of the ball-and-stick models already described can be used to represent various complex ions. The ammonia model (Figs. 4 and 24) is representative of such ions as ClO_3^- and IO_3^-, with the central sphere—the apex of the "pyramid"—representing, of course, the halogen atom. The methane model (Fig. 5) has the same shape as the following well-known species: BF_4^-, PO_4^{---}, SO_4^{--}, ClO_4^-, $CoCl_4^{--}$, $AlBr_4^-$, CrO_4^{--}, and hydrated ions such as $Be(H_2O)_4^{++}$ and $Zn(H_2O)_4^{++}$. These are all tetrahedral complexes, with the first atom in the formula being located at the center, and the other atoms, at the verticies, of a regular tetrahedron (Fig. 25).

Models representative of three other common types of complex ions can be made with the aid of the jigs shown in Fig. 30. These are made from cardboard, for use with stiff wire punches, as described on p. 49. A sphere centered in the triangular planar jig (Fig. 30a), and punched with all three punches, *A*, *B*, and *C*, can then become the central "atom" in the model of, say, the NO_3^- ion, shown in Fig. 31a. The space-filling model of this ion, shown in Fig. 31b, can be made in the usual way, by first drawing the spheres to scale on paper, and thus obtaining the distances needed to accurately mark and slice the spheres. This could just as well represent the carbonate, CO_3^{--} ion, which has the same structure.

Many complexes have the central atom bonded to six neighbors, or so-called ligands, where the former is at the center, and the latter at the vertices of, a regular octahedron (Fig. 33). To prepare a central "atom", punch it with all four punches of the octahedral jig (Figs. 30b and 32). Keeping punches *C* and *D* in place, rotate the sphere 90° about this *C–D* axis (Fig. 32b), until the holes made by punches *A* and *B* are at right angles to the plane of the jig itself. Then make the fifth and sixth holes by reinserting punches *A* and *B* into the sphere (Fig. 32c). This sphere can now serve as the central "atom" in models such as the AlF_6^{---} shown in Fig. 34. This is also the structure of such

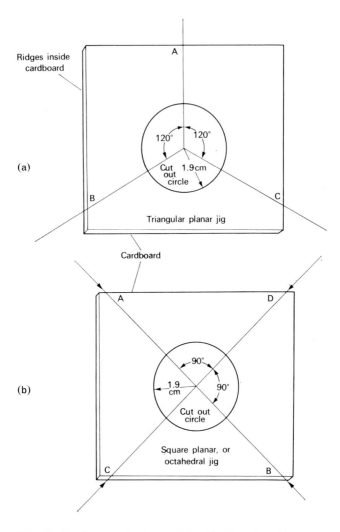

FIG. 30. Jigs for complex ion models: (a) triangular; (b) square planar and octahedral.

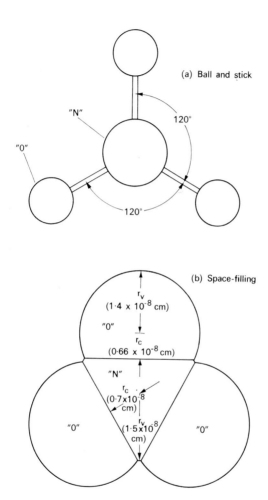

FIG. 31. (a) Ball-and-stick, (b) space-filling models of NO_3^- planar triangular.

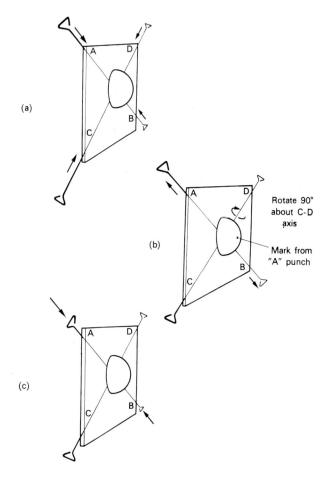

FIG. 32. (a)–(c) Use of jig in preparing octahedral complex center sphere.

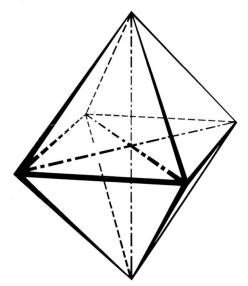

FIG. 33. Drawing of regular octahedron, showing that lines joining opposite vertices intersect at right angles, at the geometric center. This solid has eight congruent, equilateral triangles for faces, and can be pictured as two square pyramids sharing a common base.

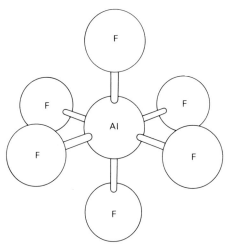

FIG. 34. Ball-and-stick model of AlF_6^{---}, typical octahedral complex.

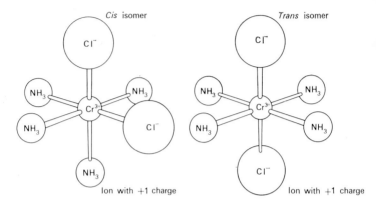

Fig. 35. Ball-and-stick models of *cis*- and *trans*-isomers of
$Cr(NH_3)_4Cl_2^+$.

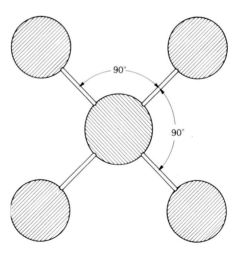

Fig. 36. Ball-and-stick model of square planar arrangement.

common species as $Fe(CN)_6^{---}$ and $Fe(CN)_6^{----}$, as well as of $Al(H_2O)_6^{+++}$, $Cr(H_2O)_6^{+++}$, $Cr(NH_3)_6^{+++}$, $Cr(NH_3)_4Cl_2^+$ and related ions. The models of such complexes are especially useful in explaining such things as the existence of the two isomers of $Cr(NH_3)_4Cl_2^+$ (Fig. 35).

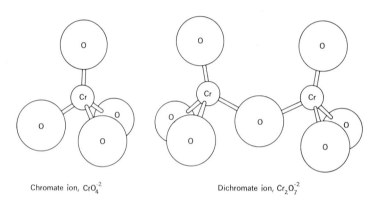

Chromate ion, CrO_4^{-2} Dichromate ion, $Cr_2O_7^{-2}$

FIG. 37. Ball-and-stick models of CrO_4^{--} and $Cr_2O_7^{--}$ ions.

The square planar arrangement, shown in Fig. 36, is the structure of complexes such as $Ni(CN)_4^{--}$, $PtCl_4^{--}$, and the famous $Cu(NH_3)_4^{++}$. Its central sphere is prepared on the Fig. 30b jig simply by punching it with all four punches, *A*, *B*, *C* and *D*.

Interested readers may wish to construct yet other models of complex ions, and are referred, for structural information, to the Appendix bibliography. Such models often show the relation of even larger complexes to those already considered. The CrO_4^{--} ion, for example, with the tetrahedral structure shown in Fig. 5, is closely related to the $Cr_2O_7^{--}$, as shown in Fig. 37. The latter may be considered two CrO_4 tetrahedra, connected by the oxygen "bridge", as shown in the figure.

ORGANIC MOLECULAR MODELS

Introduction

This unit is at once an extension of Unit 3 and Unit 5. The Unit 3 extension applies particularly to the construction and use of ball-and-stick models, and it is strongly recommended that the student organic chemistry ball-and-stick kit, previously mentioned in Units 2 and 3, be used for these models. This portion of the following material will be of interest, and understandable, to the lay reader. On the other hand, references to space-filling models, although kept purposely to a minimum, are clearly a continuation of material in Unit 5, and as such, of interest probably only to readers with chemistry backgrounds.

While the use of the commercially available ball-and-stick kit is almost a necessity in following all discussion concerning these ball-and-stick models, the space-filling models are best made from styrofoam spheres, prepared in much the same manner as has been detailed in Unit 5. Readers interested in such constructions are referred to the Appendix references for the necessary information.

Organic Chemistry: the Chemistry of Carbon Compounds

If you look at the structural formulas in Table 2, p. 28, you note that the number of dashes surrounding the symbol of an atom is the same as its valence. Since we have represented each valence bond by one dash, this is hardly surprising. Thus, each H or Cl has only one dash attached to it. Each O has two, each N three, and each C has four dashes. Note that this is the case whether the carbon is attached to four univalent atoms, as in CH_4 or CCl_4, or to two divalent atoms, as in CO_2. The important thing is that in each case the C has four valence bond dashes associated with it.

Because of this need of each carbon atom to make four bonds, carbon is strikingly different in its chemistry than are the other elements. This can be seen even when considering its structure as an element. Both hydrogen and chlorine, for example, exist in their elementary forms as **di-atomic** (two-atom) molecules. A look at their structural formulas, as given in Table 2, p. 28, shows why. In both the H_2 and Cl_2 molecules, each atom has formed all the bonds it can, namely one. The H_2, or Cl_2, unit is therefore satisfied in the sense that it has no extra bonds left over with which it might attach itself to other atoms. Now consider carbon, and try to write its structural formula as an element. To start with, a first C atom must be surrounded by four others:

$$
\begin{array}{c}
\ |\ \\
-\!C\!- \\
|\ \ \ |\ \ \ | \\
-C\!-\!C\!-\!C- \\
|\ \ \ |\ \ \ | \\
-\!C\!- \\
\ |\
\end{array}
$$

for it is known that a $C{\equiv}C$ arrangement cannot exist. Now the center C is satisfied, but the four outside ones are not because they still have bonds to complete. Well, why not add other C atoms to these latter ones? This would give:

$$
\begin{array}{c}
|\ \\
-C- \\
|\ \ |\ \ | \\
-C-C-C- \\
|\ \ |\ \ |\ \ |\ \ | \\
-C-C-C-C-C- \\
|\ \ |\ \ |\ \ |\ \ | \\
-C-C-C- \\
|\ \ |\ \ | \\
-C- \\
|
\end{array}
$$

It should now be apparent that no matter how many times you repeat the process, the outer C atoms will never be satisfied. What you have drawn is the two-dimensional analog of the diamond form of carbon.

Because of this unique valence problem, carbon has the ability to form long chains with itself. In addition, it can bond with other elements. As a result, carbon forms more compounds than all other elements put together. These carbon compounds are known as **organic compounds,** and a special branch of chemistry called **organic chemistry** has been set up to study them.

Hydrocarbons—the Simplest Organic Compounds

We have already modeled the methane molecule (Fig. 5 and Fig. 24), in which a carbon atom is bonded to four hydrogens. Carbon can also bond with itself, as we have just seen, and we might therefore just as easily make a molecule of, say, two carbons and some hydrogens. From a commercially available ball-and-stick kit, join two black spheres, representing carbon atoms, using a long dowel to bond them, and insert short dowels in all the other holes. This would give

$$\begin{array}{c} \quad | \quad | \\ -C-C- \\ \quad | \quad | \end{array}$$

to which we must attach six hydrogen atoms in order that there be no valence bonds left over. This gives the well known ethane molecule (Fig. 38), similar in properties to the gas methane, and having the molecular formula C_2H_6, and the structural formula

$$\begin{array}{c} H \quad H \\ | \quad | \\ H-C-C-H \\ | \quad | \\ H \quad H \end{array}$$

There is no reason to stop here, for we could just as well build a molecule with three carbon atoms in the same way, giving

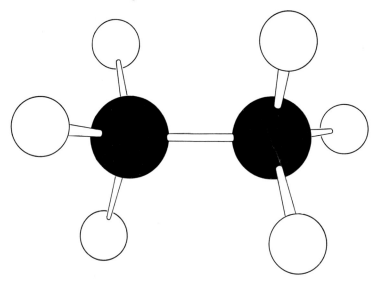

FIG. 38. Model of ethane molecule, C_2H_6.

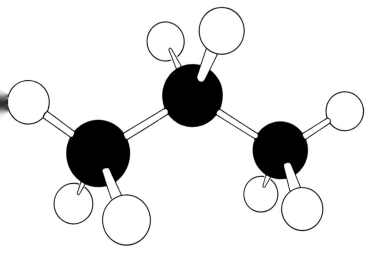

FIG. 39. Model of propane, with molecular formula C_3H_8.

Here we would attach eight H atoms to fill all bonds, and end with a molecule of propane (Fig. 39), the other principal component of natural gas. Its formula would be C_3H_8, or, better still,

```
      H   H   H
      |   |   |
  H—C—C—C—H
      |   |   |
      H   H   H
```

SPACE-FILLING MODELS OF HYDROCARBONS

Although we shall not attempt to describe the construction of accurate space-filling models of the organic compounds considered on the following pages, the interested reader is reminded that such constructions are possible from styrofoam spheres, and that the general methods have already been detailed in Unit 5. It is appropriate at this point to perhaps use the ethane molecule, whose ball-and-stick model is shown in Fig. 38, as an example of these space-filling model constructions.

Essentially, the spheres are prepared as detailed on p. 53 and Fig. 23. The difference is that only three small, 2.4-cm radius spheres, representing the hydrogen atoms, are joined to each larger, 3.4-cm radius (carbon atom) spheres, so that in this sense, each large sphere resembles the ammonia model (Fig. 23). The fourth "bond" for each "carbon atom" is to the other "carbon", and this must be considered when preparing the spheres.

In practice, one prepares all six "hydrogen atoms" from the 2.4-cm diameter spheres exactly as described on pp. 47–49 and Fig. 18. The two larger spheres are each prepared as shown in Fig. 40. First, two-dimensional circles, of the same radius as the spheres, are drawn, and angle AOC is measured off (Fig. 40a) to determine the distance, b, between the centers of circles to be drawn on the spheres themselves. The locations of these centers are found as indicated on p. 53 and Fig. 23. The radius of each circle is found by measuring off the covalent radius of

0.77×10^{-8} cm (1.54 cm on our scale), OD in Fig. 40b, and constructing EF perpendicular to AB at D (Fig. 40b). Distance AF (C on Fig. 40b) is then the desired radius.

Each of the two large spheres then has the four circles, each with radius "c", and whose centers are separated by distance "b",

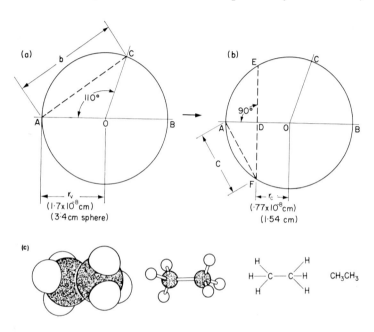

FIG. 40. (a) and (b) Space-filling model of ethane. Preparing "carbon atoms"; (c) structural formulas of ethane, C_2H_6.

drawn on it with compass, and then sliced off in the usual manner. The spheres are then joined, the smaller spheres added, to give the model shown at left of Fig. 40c. Figures 40c and 41 contrast the appearance of the different models, and formula representations, of ethane as well as several five-carbon hydrocarbons.

Perhaps you can already see why there are so many organic compounds. Even if we consider only compounds of hydrogen and carbon, appropriately known as **hydrocarbons**, there is no reason why we cannot go on and construct molecules of four, five, six or more carbon atoms in length. Indeed, hydrocarbons of as many as fifty carbon atoms per molecule are known to exist, and have also been made in the laboratory by organic chemists.

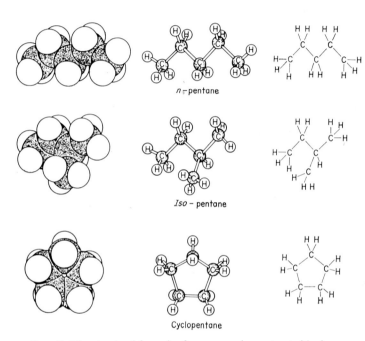

FIG. 41. The structural formulas for some carbon saturated hydrocarbons. Contrasting the appearance of space-filling (left) and ball-and-stick (center) models of compounds whose structural formulas are given at right.

Each hydrocarbon we have thus far considered has had only single bonds. Each bond that did not join two carbon atoms together was therefore free to attach itself to a hydrogen atom. Such single-bonded hydrocarbons are called saturated because they have as many hydrogen atoms in their molecules as is possible (i.e. are saturated with them). To emphasize this, you might build some double-bond hydrocarbons, using the bendable metal springs included in the ball-and-stick kit for double bonds. With two carbon atoms, for example, you would find that once you connected them with the two-spring double bond, giving

$$
\begin{array}{cc}
| & | \\
C & = C \\
| & |
\end{array}
$$

there would be room for only four hydrogen atoms, and you would end with

$$
\begin{array}{cc}
H & H \\
| & | \\
C & = C \\
| & | \\
H & H
\end{array}
$$

the gaseous compound ethylene (Fig. 42), also known as ethene. Do not forget that the bond between two atoms belongs to both atoms, so that here, too, each carbon has four valence bonds, and each hydrogen, one.

As you may have guessed, there is a whole series of these double-bonded hydrocarbons. To build the next member of this series, connect two black spheres with two of the metal springs,

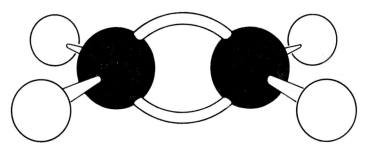

FIG. 42. Model of ethylene, C_2H_4, showing double bond.

to represent the double bond, and use a long dowel rod to connect
a third black sphere to the other two. This gives

$$
\begin{array}{c}
\mid \quad\; \mid \quad\; \mid \\
C{=}C{-}C{-} \\
\mid \quad\quad\; \mid
\end{array}
$$

to which six white spheres would then be attached by six short
dowels, giving the propylene, or propene molecule (Fig. 43),
C_3H_6, of structural formula

$$
\begin{array}{c}
H \quad H \quad H \\
\mid \quad\; \mid \quad\; \mid \\
C{=}C{-}C{-}H \\
\mid \quad\quad\; \mid \\
H \quad\quad\; H
\end{array}
$$

Since two bonds are used to connect the first two C atoms, less
are available for hydrogen atoms. Hence these molecules contain
less than the maximum, or **saturated**, number of H atoms, and
are therefore named unsaturated hydrocarbons. Perhaps you
have come across the terms "saturated" and "unsaturated",
used to describe various cooking fats. The saturated fats are
those which contain only single carbon-to-carbon bonds, while
the unsaturated ones have one or more double bonds in their
molecules. Models of simple unsaturated hydrocarbons are
shown in Fig. 43.

   Although less common than the already discussed hydrocarbons,
yet another group of them is known in which three valence
bonds hold together two carbon atoms, forming a so-called triple
bond. We shall make only the first member of this series, the gas
acetylene, by connecting two black spheres with three of the
metal springs, thereby forming the triple bond, and giving
$-C{\equiv}C-$. Because this has used up three of each carbon's four
bonds, only one per atom is still available to form a bond with
hydrogen. Hence only two white spheres can be connected with

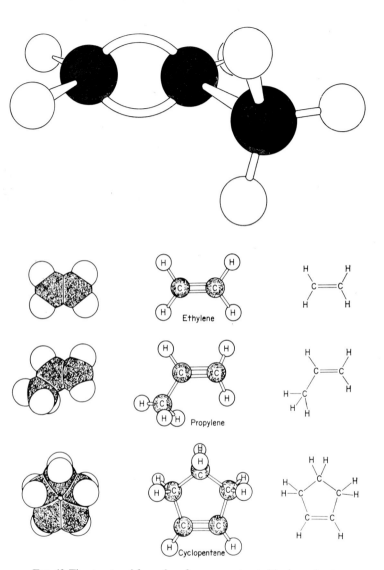

FIG. 43. The structural formulas of some unsaturated hydrocarbons. The relationship of propylene, C_3H_6, to ethylene becomes apparent if you think of it as ethylene, in which one H atom has been replaced by a —CH_3 group.

short dowels to the remaining holes, giving the acetylene molecule C_2H_2 (Fig. 44), of structural formula $H—C \equiv C—H$.

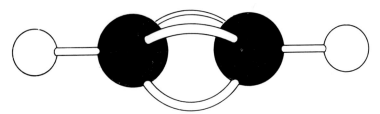

FIG. 44. Model of acetylene, C_2H_2, showing the triple bond between the carbon atoms.

STRUCTURAL ISOMERS

Besides the three different types of hydrocarbons just discussed, there is another more important reason for the vast number of organic compounds. To understand it, construct a four-carbon hydrocarbon in the usual way, using long dowels to connect the black spheres, and short ones to all white spheres. The result should look like Fig. 45, and represents a molecule of normal butane, C_4H_{10}, with the structural formula

$$
\begin{array}{cccc}
\text{H} & \text{H} & \text{H} & \text{H} \\
| & | & | & | \\
\text{H—C} & \text{—C} & \text{—C} & \text{—C—H} \\
| & | & | & | \\
\text{H} & \text{H} & \text{H} & \text{H}
\end{array}
$$

With a piece of white chalk write the number "1" on the first black sphere, a "2" on the second, and so on, to correspond to

$$
\begin{array}{cccc}
\text{H} & \text{H} & \text{H} & \text{H} \\
| & | & | & | \\
\text{H—C}_1 & \text{—C}_2 & \text{—C}_3 & \text{—C}_4\text{—H} \\
| & | & | & | \\
\text{H} & \text{H} & \text{H} & \text{H}
\end{array}
$$

This straight chain normal butane is a gas at room temperature, but it condenses to a liquid at 0°C (32°F), the temperature at which water freezes.

Now there is another compound, obtained from coal tar, which has been shown by analysis also to have the molecular formula C_4H_{10}. It is therefore butane, but strangely enough its properties differ slightly from those of normal butane. To condense it, for example, you would have to cool it some 15° below the temperature at which normal butane condenses. There is only one way to explain this difference, and that is by assuming slight differences

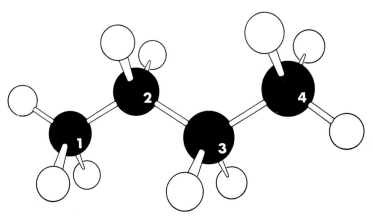

FIG. 45. Model of normal butane, molecular formula C_4H_{10}, a straight chain hydrocarbon.

in the shapes of the molecules themselves. If you examine your model of normal butane (Fig. 45), you will see still another way of forming it. Remove carbon number 4, with its long dowel and three connected H atoms, from carbon 3. At the same time, remove one H atom, with its dowel, from carbon 2. The result so far is shown in Fig. 46a. In two dimensions, we can picture it thus:

$$
\begin{array}{ccc}
 & H & \\
H & | & H \\
| & | & | \\
H-C_1-C_2-C_3 \\
| & | & | \\
H & H & H
\end{array}
\qquad
\begin{array}{c}
H \\
| \\
-C_4-H \\
| \\
H
\end{array}
$$

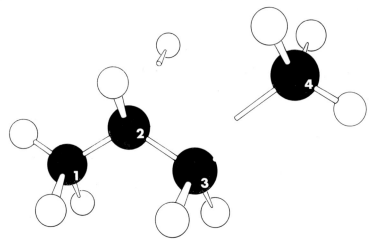

FIG. 46a. Showing how to change normal butane into iso-butane.
After the H atom and the —CH₃ group are removed, their positions
are switched.

Now switch positions, attaching the removed H onto the third
carbon, and number 4 carbon (with its three H atoms) to number
2 carbon. The molecule thus becomes

$$
\begin{array}{c}
\mathrm{H} \\
| \\
\mathrm{H-\underset{4}{C}-H} \\
\mathrm{H} \quad\quad \mathrm{H} \\
| \quad\quad | \\
\mathrm{H-\underset{1}{C}-\underset{2}{C}-\underset{3}{C}-H} \\
| \quad | \quad | \\
\mathrm{H \quad H \quad H}
\end{array}
$$

and should resemble Fig. 46b. Note that while it also has the
formula C_4H_{10}, and satisfies all valence requirements, it does have
a different molecular shape. This difference is due to the replacing
of the straight chain of four carbon atoms

$$
\begin{array}{c}
| \quad | \quad | \quad | \\
\mathrm{—C—C—C—C—} \\
| \quad | \quad | \quad |
\end{array}
$$

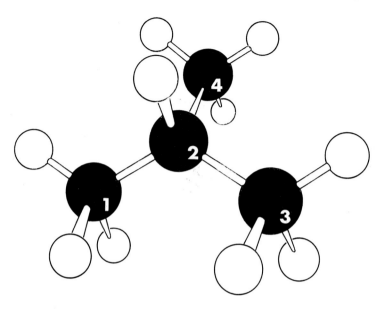

Fig. 46b. The result of the switching indicated in Fig. 46a, isobutane. Note that while its formula is the same as that of normal butane, it has a different structure.

by the branched chain,

$$-\overset{|}{\underset{|}{C}}-$$
$$-\overset{|}{\underset{|}{C}}-\overset{|}{\underset{|}{C}}-\overset{|}{\underset{|}{C}}-$$

This latter molecule is called iso-butane (Fig. 46b), to distinguish it from normal butane (Fig. 45). The two molecules are known as **isomers** because they have the same molecular formula, yet differ in structure. This difference in structure must be what accounts for the difference in properties of the two compounds.

Now try the same kind of rearranging with the five-carbon

chain compound, pentane, formula C_5H_{12}. First, build normal pentane (Fig. 47) in the usual way, numbering the black spheres

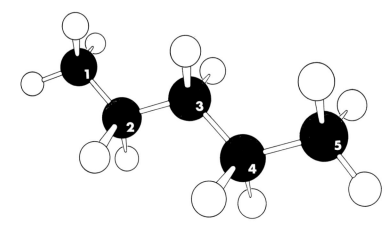

FIG. 47. Model of normal pentane, C_5H_{12}.

consecutively, 1 through 5. The structural formula is

$$
\begin{array}{ccccc}
H & H & H & H & H \\
| & | & | & | & | \\
H-C_1-C_2-C_3-C_4-C_5-H \\
| & | & | & | & | \\
H & H & H & H & H
\end{array}
$$

or, for simplicity

$$
\begin{array}{ccccc}
| & | & | & | & | \\
-C_1-C_2-C_3-C_4-C_5- \\
| & | & | & | & |
\end{array}
$$

omitting the H atoms for convenience, but remembering that one belongs to the end of each unattached dash. Now, by switching H atoms and C atoms with their attached hydrogens, see how many different isomers of pentane you can build. Remember, they must each have the molecular formula C_5H_{12}, and have all their valence bonds completed. You should have found two additional possible isomers, as shown in Fig. 48, with the structural formulas

```
           |
          -C-
          |5
  |    |    |    |
 -C — C — C — C-
  |1   |2   |3   |4
```

and

```
          |
         -C-
         |5
  |    |    |
 -C — C — C-
  |1   |2   |3
         |
        -C-
         |4
```

Again, for the sake of simplicity, H atoms are not shown; but again you must imagine one attached to the end of every dash, except those between C atoms. It turns out that chemists have found that three different isomers of pentane do indeed exist. They correspond to the normal straight chain pentane (Fig. 47), and to the two-branched chain isomers (Fig. 48) whose structural formulas are given above.

Do not make the mistake of thinking that there should be other pentane isomers just because you can write structures like

```
               |
              -C-
  |    |    |  |
 -C — C — C — C-
  |    |    |  |
```

or

```
          |
         -C-
  |    |  |
 -C — C — C-
  |    |  |
         -C-
          |
```

These are tricks of our two-dimensional paper, for the first is identical to normal pentane when you model it, while the latter is just like the first isomer shown above. The surest way to

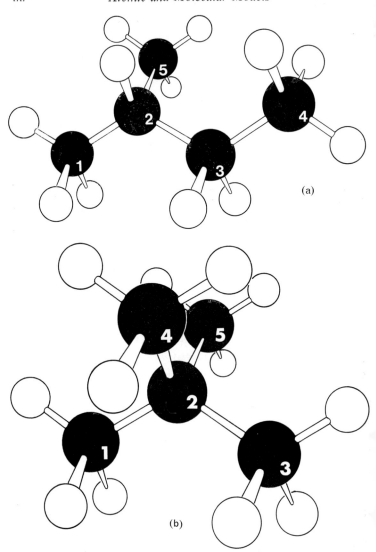

(a)

(b)

Fig. 48 (a) and (b). Models of the two iso-pentanes, made by switch-ing first one and then the other of carbon 2's H atoms with the —CH₃ groups on the far right of Figs. 47 and 48a, respectively.

avoid such mistakes on paper is to number the C atoms. Then you will see that

$$-\overset{|}{\underset{|}{C}}_1 - \overset{|}{\underset{|}{C}}_2 - \overset{|}{\underset{|}{C}}_3 - \overset{|}{\underset{|}{C}}_4 - \overset{|}{\underset{|}{C}}_5 -$$

and

$$\begin{array}{c} -\overset{|}{\underset{|}{C}}_5 - \\ -\overset{|}{\underset{|}{C}}_1 - \overset{|}{\underset{|}{C}}_2 - \overset{|}{\underset{|}{C}}_3 - \overset{|}{\underset{|}{C}}_4 - \end{array}$$

each have C_5 bonded to C_4, while in both

$$\begin{array}{c} -\overset{|}{\underset{|}{C}}_5 - \\ -\overset{|}{\underset{|}{C}}_1 - \overset{|}{\underset{|}{C}}_2 - \overset{|}{}{C}_3 - \overset{|}{\underset{|}{C}}_4 - \end{array}$$

and

$$\begin{array}{c} -\overset{|}{\underset{|}{C}}_5 - \\ -\overset{|}{\underset{|}{C}}_1 - \overset{|}{\underset{|}{C}}_2 - \overset{|}{}{C}_3 - \\ -\overset{|}{\underset{|}{C}}_4 - \end{array}$$

the C_4 is joined to C_3. It is only when a given C can be joined to different C atoms that separate isomers actually exist.

A ball-and-stick kit will probably have enough black and white spheres to allow construction of six, seven, and eight-chain C hydrocarbons; or, each one, in turn, of their various isomers. You may be amazed to learn the number of different isomers that can be made, and which are actually known to exist, for each of these hydrocarbons. For a six carbon chain, the number is five; for a seven carbon chain it jumps to nine, and for an eight carbon saturated hydrocarbon, there are eighteen possible isomers. You may wish to consult an organic chemistry test (see Appendix bibliography) to check your models, if you feel inclined to make them. Now perhaps you can see why so many different organic compounds are known, just among the hydrocarbons. Thirty-five different isomers with the structural formula C_9H_{20} are known, and for the hydrocarbon $C_{20}H_{42}$, over 100,000 different isomers are theoretically possible!

CARBON GEOMETRY AND GEOMETRIC ISOMERISM

You have surely noticed that the holes in each black sphere in the ball-and-stick kit were drilled so that dowel bonds inserted in them point outward at certain definite angles. Why were these angles chosen, rather than some other set? Let us make use of our models in an attempt to find an answer.

To begin with, consider the methane molecule (Fig. 5), and note that all four H atoms are equivalent. This means that if you replace an H (white sphere) by a Cl atom (green sphere), you will get the same result no matter which H you use. Try it with your methane model to convince yourself that this is so. Now, the question arises, is there any other methane model which would give the same result? The answer is that there are two others (Fig. 49), known as square planar and pyramidal because of their shapes. Note that in both cases, all H atoms are equivalent. Why, you may ask, should a model of methane have all its H atoms equivalent anyway? Because chemists actually have replaced one of methane's H atoms with a Cl atom, to give the compound

known as methyl chloride. Its name comes from the

group of atoms it contains, which is called a methyl group. Although this compound has been made many times by many

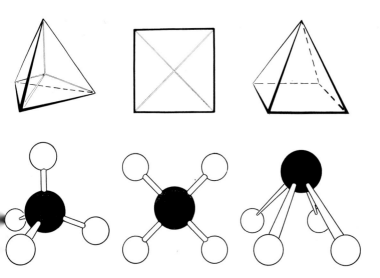

FIG. 49a. Three possible models of methane, CH₄, in which all four H atoms are equivalent. From left to right, they are tetrahedral, planar, and pyramidal. The names come from their resemblance to the tetrahedron, square plane, and pyramid, shown above them.

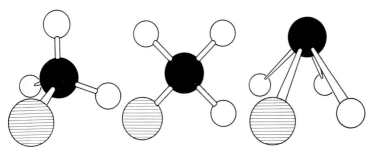

FIG. 49b. Three possible models of methyl chloride, CH₃Cl. Note that, in each case, it does not matter which H has been replaced by a Cl atom. Hence, all H atoms of each model are said to be equivalent.

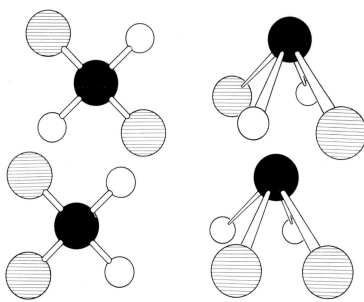

Fig. 50. Planar (left) and pyramidal models of CH_2Cl_2, showing two possible arrangements for each. The top models have the Cl atoms opposite each other, while in the lower models they are adjacent.

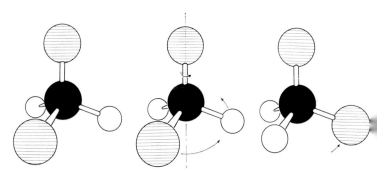

Fig. 51. Tetrahedral model of CH_2Cl_2, showing that if the model on the left is rotated about its vertical axis (center), it looks like the model on the right. Thus, only one CH_2Cl_2 can be constructed from a tetrahedral model.

people, only one kind of methyl chloride, with one set of properties, has ever been found. It therefore follows that no matter
which of the four H atoms was replaced, the same compound
must have resulted. Hence the necessity of the equivalence of all
hydrogens.

All this shows only that any one of the three models—in
Fig. 49a—are equally acceptable. Why did we choose the former?
To see, replace a second H of your methane with a second Cl
atom, and imagine doing the same thing to each model of Fig. 49b.
The results are shown in Fig. 50. The significant thing for us is
that according to either the planar or pyramidal models, two
isomers of this

$$
\begin{array}{c}
H \\
| \\
H-C-Cl \\
| \\
Cl
\end{array}
$$

compound are possible. Only our methane model, called tetrahedral because its bonds extend to the corners of a tetrahedron
(Fig. 51), shows that there is only one

$$
\begin{array}{c}
H \\
| \\
H-C-Cl \\
| \\
Cl
\end{array}
$$

no matter which two H atoms you replace. Interestingly enough,
only one kind of

$$
\begin{array}{c}
H \\
| \\
H-C-Cl \\
| \\
Cl
\end{array}
$$

has ever been found, and this is why chemists accept our tetrahedral model of methane.

Let us carry this kind of reasoning a step further, and apply it to the ethane model of Fig. 38, and to the double-bonded ethylene of Fig. 42. Note first, that no matter which of the six H atoms of ethane you replace with a Cl, the result is the same. This is also true of the four hydrogens of ethylene, and these predictions check with experiment—even though it is very difficult to prepare the compound

```
        H       Cl
        |       |
        C ===== C
        |       |
        H       H
```

In other words, there is only one

```
          H       Cl
          |       |
    H—C————C—H
          |       |
          H       H
```

and only one kind of

```
        H       Cl
        |       |
        C ===== C
        |       |
        H       H
```

known, and this is in agreement with the models you have made

Now replace one of the other carbon's H atoms on each model with a second Cl atom. In the case of the saturated hydrocarbon it makes no difference where the second Cl is attached, because

```
          Cl      H
          |       |
    H—C————C—H
          |       |
          H       Cl
```

becomes

```
       Cl    Cl
       |     |
 H—C——————C—H
       |     |
       H     H
```

imply by twisting the C atoms around their single dowel rod bond (see Fig. 52). Experiments have shown that only one

```
       Cl    H
       |     |
 H—C——————C—H
       |     |
       H     Cl
```

does, in fact, exist. What about the double-bonded compound?

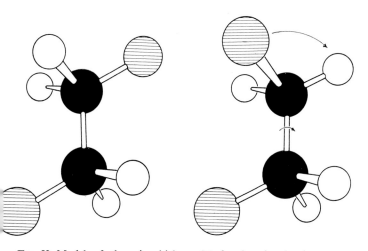

FIG. 52. Models of ethane in which one H of each carbon has been replaced by a Cl atom. By rotating model at left about its vertical axis (see arrows), the two Cl atoms can be switched from the same side to opposite sides (right).

Here, there are two distinct possibilities:

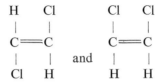

$$
\begin{array}{ccc}
\text{H} \quad\ \text{Cl} & & \text{Cl} \quad \text{Cl} \\
| \qquad | & & | \qquad | \\
\text{C}=\!\!=\text{C} & \text{and} & \text{C}=\!\!=\text{C} \\
| \qquad | & & | \qquad | \\
\text{Cl} \quad\ \text{H} & & \text{H} \quad \text{H}
\end{array}
$$

(see Fig. 53a), because no amount of twisting will make them identical—it will only break the double bond. Chemists have actually found that two separate, stable $C_2H_2Cl_2$ isomers, with slightly different sets of properties, do exist.

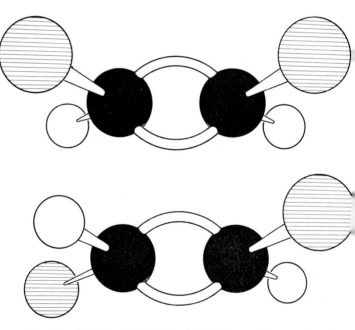

FIG. 53a. Models of double-bonded ethylene with one of each carbon's H atoms replaced by Cl atoms show that now the Cl atoms can be placed either on the same side, or on opposite sides of the double bond. Here, one can NOT be twisted to give the other, without breaking the double bond.

The

is called the *cis*, and

$$
\begin{array}{cc}
\text{Cl} & \text{H} \\
| & | \\
\text{C} =\!\!= \text{C} \\
| & | \\
\text{H} & \text{Cl}
\end{array}
$$

the *trans* isomer. Figure 53b contrasts ball-and-stick with space-filling models. A third, less-stable, possibility is shown at the top of Fig. 53b. These facts give us increased confidence in the validity of our models.

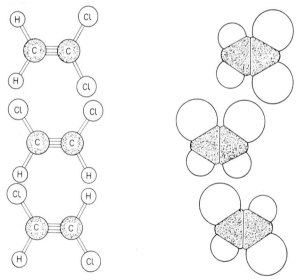

FIG. 53b. The isomers of dichloroethylene. Contrasting ball-and-stick with space-filling models of the $C_2H_2Cl_2$ isomers.

Besides the types we have so far considered, there is another, and more subtle, difference that can exist between two compounds with the same molecular formula. To understand the nature of this difference, hold your right hand up to a mirror. The reflection looks the same as your left hand, a fact which becomes apparent if you hold both hands in front of you, with fingers pointing up and thumbs toward you. Now imagine a mirror placed between your hands, facing the right hand, and you will see that your left hand does indeed correspond with the mirror image of your right hand. In other words, your hands are mirror images of each other. The isomers we now want to talk about are also mirror images of each other, and are called **stereoisomers**.

Stereoisomerism can occur only among molecules where a central carbon atom is bonded to four different kinds of atoms, or groups of atoms. Such carbon atoms are called asymmetric carbons because there is no symmetry about their neighbors, and you can model one as follows: insert one short and three long dowels into each of two black spheres. Attach white spheres to each short dowel, and then attach a green, a red, and a blue sphere to the long dowels of one of the black spheres. This black sphere now has four different spheres attached to it, and therefore represents an asymmetric carbon atom. Now take the second black sphere, and attach to it a green, a red and a blue sphere, but in such a way that it becomes the mirror image of the first asymmetric carbon model (Fig. 54). It may help you to hold this first model up to a mirror as you construct the second one. The two mirror-image models are stereoisomers, but they are not identical. There is no way that you can make one correspond exactly to the other, no matter how you turn them. Try it and see.

Because of their similarity, you might expect only very slight differences in properties between such isomers. This is indeed the case, for they differ in only one way: in their effect on light,

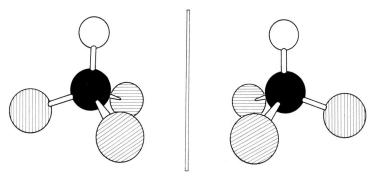

Fig. 54. Models of simple stereoisomers. To show that they are mirror-images, imagine a mirror (the vertical line represents its edge) placed to the right of the model shown at left. Its mirror-image would be the model shown at right.

and on light that has been polarized. This is explained in the references listed at the end of the book.

This kind of behavior was observed over 100 years ago by Louis Pasteur, in his study of tartaric acid, $C_4H_4O_6$. To explain this behavior, let us model a similar, but simpler compound, lactic acid, $C_3H_6O_3$, with the structural formula

$$
\begin{array}{ccccccc}
 & H & & H & & O & \\
 & | & & | & & \| & \\
H\text{---}C & \text{------} & C & \text{------} & C & \text{------} & O\text{---}H \\
 & | & & | & & & \\
 & H & & O\text{---}H & & &
\end{array}
$$

for its model shows just what tartaric acid would—and avoids the confusion of a more complicated molecule. Build two models of lactic acid, employing the usual conventions regarding dowels and spheres, and using metal springs for the double bonds. When you have done this, you will see that the middle C atom of each model is asymmetric, because it has four different atoms, or groups of atoms, attached to it.

These are an H atom, an O—H group, a H—C— group and a

$$\begin{array}{c} H \\ | \\ H-C- \\ | \\ H \end{array}$$

O
‖
—C——O—H group.

This means that stereoisomers are possible, and you might therefore try changing the positions of these groups around the middle carbon of one model, until it becomes a mirror image of the other. A set of such models is shown in Fig. 55.

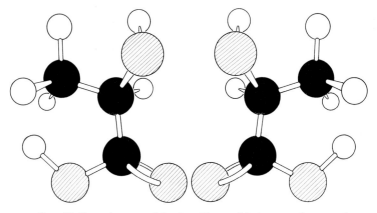

FIG. 55. Stereoisomers of lactic acid, possible because the central carbon atom has four different groups attached to it. Again imagine a mirror between the two models: one can be seen to be the mirror-image of the other.

To illustrate the importance of asymmetry, consider the compound propionic acid, $C_3H_6O_2$, with structural formula

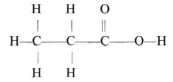

differing from lactic acid only in that an H has replaced the O—H on the middle carbon. Yet this slight change has been enough to make that middle carbon asymmetric no longer, because it now has two identical H atoms attached to it. If you replace each O—H of your lactic acid models with H atoms, they become

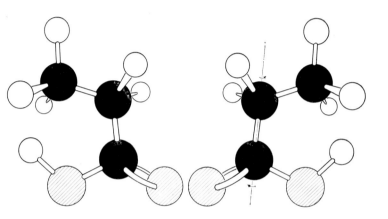

Fig. 56. Mirror-image models of proprionic acid, identical with the lactic acid models of Fig. 55 except that the OH groups attached to the central carbons have been replaced by H atoms. Because of this, the central carbons no longer have four DIFFERENT groups attached to them, and are hence no longer asymmetric. The result of this be- comes apparent if the bottom part of the model on the right is twisted to the left, while the top part is twisted to the right, about a vertical axis (see arrows).

models of propionic acid molecules. While it is true that you can make these mirror images of each other (Fig. 56), it is also true that by simply rotating the double bond C about the dowel connecting it with the middle carbon, the molecules can be made to coincide (Fig. 57). Thus they are not really isomers at all, but identical molecules. The lactic acid models could not be made to coincide, no matter how they were rotated, and they therefore are true stereoisomers, just as the more complicated tartaric acid molecules are. While only one kind of propionic acid, with one set of properties, is known, both lactic and tartaric acid do have isomers which are identical except in their effect on light. Your models, as you have seen, explain why.

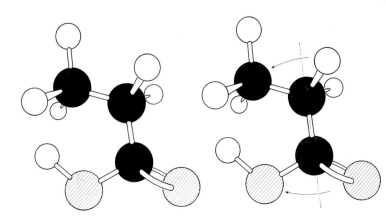

Fig. 57. The models of Fig. 56, but after the one at right has been twisted as indicated in Fig. 56. Note that the two models are now identical, and no longer mirror-images.

Different Types of Organic Compounds

Except for the acid models just discussed, you have modeled only hydrocarbons, and a few of their chlorine compounds. Because of carbon's ability to bond with other atoms besides hydrogen, chlorine and itself, many other organic compounds are known, and still others have been made by organic chemists in the laboratory. The materials in most ball-and-stick kits enable you to build models of many of these important compounds. We will now give you the structural formulas of some of these, so that you may model them, if you wish. We will also tell you a little about the more widely known ones.

Remember the conventions about long and short dowels, the use of metal springs to represent double bonds, and the color code we have used. Remember also that what you build are models of these compounds' molecules; models that are accepted and used by chemists because they explain much about the behavior of these compounds, as we have already seen. But they are not meant to be pictures of these molecules—simply useful models of them, which is quite a different thing.

A majority of the following compounds may be considered to be derived from the hydrocarbons we have already spoken of. This is not necessarily because they may be made from hydrocarbons, although this is sometimes the case, but rather because of the close structural resemblance between a given hydrocarbon and its so-called derivative. As you will see, this difference is often simply the replacing of one H atom by a group of other atoms. Because of this, it is reasonable to name such compounds so as to reveal these resemblances. We have already introduced this idea when we considered the compounds

$$
\begin{array}{ccc}
\quad\ \text{H} & & \quad\ \text{H} \\
\quad\ | & & \quad\ | \\
\text{H}-\text{C}-\text{H} & \text{and} & \text{H}-\text{C}-\text{Cl} \\
\quad\ | & & \quad\ | \\
\quad\ \text{H} & & \quad\ \text{H}
\end{array}
$$

both of which contain the

$$
\begin{array}{c}
\text{H} \\
| \\
\text{H}-\text{C}-\ \text{group} \\
| \\
\text{H}
\end{array}
$$

or radical, as it is often called. Because of its relation to methane, it is known as the methyl radical.

```
        H
        |
    H—C—Cl
        |
        H
```

is therefore named methyl chloride. By the same reasoning, since

```
    H  H                    H  H
    |  |                    |  |
H—C—C—H is ethane,  H—C—C— is an ethyl radical
    |  |                    |  |
    H  H                    H  H
```

```
        H  H
        |  |
  and H—C—C—Cl would be ethyl chloride.
        |  |
        H  H
```

ALCOHOLS

These compounds result from the replacing of an H atom of a hydrocarbon with an O—H group, known as the alcohol group. Thus

```
        H
        |
    H—C—O—H
        |
        H
```

is commonly called methyl alcohol (Fig. 58). The alcohol in liquors has the structure

```
        H  H
        |  |
    H—C—C—O—H
        |  |
        H  H
```

Methane, CH_4 Methanol, CH_3OH

FIG. 58. Structural formulas of methane and methanol, showing the relation of methyl alcohol (right) to methane (left).

and is therefore ethyl alcohol. The next in this series would be

$$
\begin{array}{ccc}
H & H & H \\
| & | & | \\
H\!-\!C\!-\!C\!-\!C\!-\!O\!-\!H \\
| & | & | \\
H & H & H
\end{array}
$$

which, because of its relation to propane,

$$
\begin{array}{ccc}
H & H & H \\
| & | & | \\
H\!-\!C\!-\!C\!-\!C\!-\!H \\
| & | & | \\
H & H & H
\end{array}
$$

is known as propyl alcohol, or I-propanol. But we might also have replaced the H of the middle carbon, thus:

$$
\begin{array}{c}
H \\
| \\
O \\
H\;\;|\;\;H \\
| \;\; | \;\; | \\
H\!-\!C\!-\!C\!-\!C\!-\!H \\
| \;\; | \;\; | \\
H \;\; H \;\; H
\end{array}
$$

Since both of these alcohols have the formula C_3H_8O, they are isomers, and the latter molecule is known as iso-propyl alcohol, or 2-propanol (Fig. 59). We shall use the common names for these, as well as the compounds that follow. Chemists, for reasons which will become apparent to you if you study organic chemistry, have found it more convenient to name all compounds in a more systematic way. You will find their system described in the references listed at the end of this book.

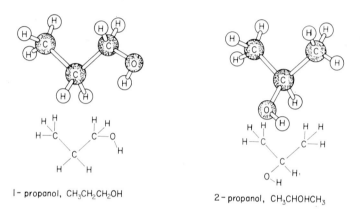

1 - propanol, $CH_3CH_2CH_2OH$

2 - propanol, $CH_3CHOHCH_3$

FIG. 59. Models of the two propyl alcohol isomers.

ALDEHYDES

Here, changes are more drastic than for alcohols, and hence the names are different. The simplest aldehyde, formaldehyde, has the structure

$$\begin{array}{c} O \\ \parallel \\ H-C-H \end{array}$$

and is used as a preservative in biology work. Note here that the difference between this compound and methane is in the replacing of two H atoms by a double-bonded oxygen. Note also the fulfilling of valence requirements: one bond for each H, two for the O, and four for each C. Remember that the double bond is shared by both atoms, and is therefore two bonds for the C as well as for the O. The next members of the aldehyde series are

acetaldehyde,

$$\begin{array}{cc} H & O \\ | & \| \\ H-C-C-H, \\ | \\ H \end{array}$$

and propionaldehyde,

$$\begin{array}{ccc} H & H & O \\ | & | & \| \\ H-C-C-C-H \\ | & | \\ H & H \end{array}$$

The thing that therefore makes a compound an aldehyde is the

$$\begin{array}{c} O \\ \| \\ C-H \text{ group,} \end{array}$$

known as the aldehyde group (see Fig. 60).

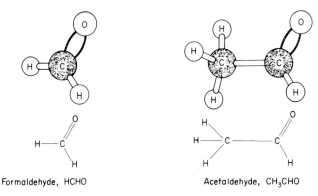

Formaldehyde, HCHO   Acetaldehyde, CH₃CHO

FIG. 60. Models of simple aldehydes.

CARBOXY ACIDS

These compounds are closely related to aldehydes, and this fact is reflected in their common names. The simplest of them is formic acid, found in ants, and having the structure

$$\begin{matrix} & O \\ & \| \\ H-&C-O-H \end{matrix}$$

Others in the series include acetic acid,

$$\begin{matrix} H & O \\ | & \| \\ H-C-&C-O-H \\ | \\ H \end{matrix}$$

better known as vinegar, and propionic acid,

$$\begin{matrix} H & H & O \\ | & | & \| \\ H-C-C-&C-O-H \\ | & | \\ H & H \end{matrix}$$

whose structure has already been considered in the experiment on stereoisomers. Note that the

$$\begin{matrix} O \\ \| \\ C-O-H \end{matrix}$$

called the carboxy group, is what makes them acids, and that this differs from the

$$\begin{matrix} O \\ \| \\ C-H \end{matrix}$$ aldehyde group

only by the extra O atom between the C and H atoms (see Fig. 61).

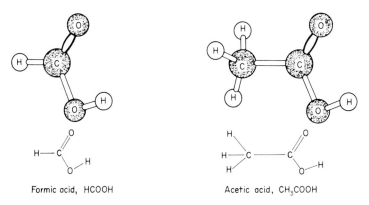

Formic acid, HCOOH   Acetic acid, CH₃COOH

FIG. 61. Models of simple carboxy acids.

KETONES

Another class of compounds closely resembling aldehydes are known as ketones. The simplest of these is the well-known solvent acetone,

$$
\begin{array}{ccc}
\text{H} & \text{O} & \text{H} \\
| & \| & | \\
\text{H—C—C—C—H} \\
| & & | \\
\text{H} & & \text{H}
\end{array}
$$

also named di-methyl ketone, for obvious reasons (see Fig. 62).

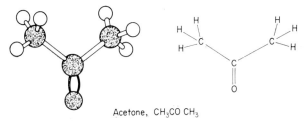

Acetone, CH₃CO CH₃

FIG. 62. Ball-and-stick model of acetone, the simplest ketone,

The name acetone comes from its resemblance to acetic acid. The next in the series,

```
        H   O   H   H
        |   ‖   |   |
   H—C—C—C—C—H
        |       |   |
        H       H   H
```

methyl ethyl ketone, shows the

```
        O
        ‖
        C
```

or keto-group, to be the distinguishing feature of these compounds. Note that while an aldehyde has a hydrocarbon radical on one side, and a hydrogen atom on the other side of this group, the ketone has hydrocarbon radicals on both sides of it.

ETHERS

Still another series of compounds connects two hydrocarbon radicals together to form a single molecule. Whereas in ketones the link is

```
        O
        ‖
        C
```

it is simply an O atom in the ethers. Di-methyl ether, for example, has the structure

```
        H       H
        |       |
   H—C—O—C—H
        |       |
        H       H
```

A close relative,

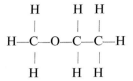

is known as methyl ethyl ether. Di-ethyl ether, the common ether used in hospitals as an anaesthetic, has the structure

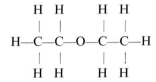

It is interesting to note that ethers have the same molecular formulas as their corresponding alcohols, and may therefore be considered isomeric with them. Ethyl alcohol, for example, and di-methyl ether both have the formula C_2H_6O (Fig. 63), but here, because of the marked difference in structure, their properties are also very different.

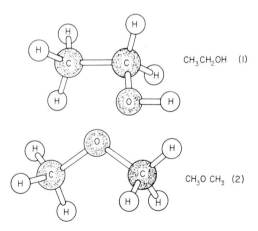

FIG. 63. Ball-and-stick models of the C_2H_6O isomers.

ESTERS

By removing the O—H of a carboxy acid, and the —H of an alcohol, it is possible to form still another class of compounds known as esters. Ethyl alcohol and acetic acid, for example, can form the ester known as ethyl acetate. Here, starting with

$$
\begin{array}{cc}
\underset{\substack{|\\ H}}{\overset{\substack{H\quad H\\ |\quad |}}{H-C-C-O-H}} & \text{and} \quad \underset{\substack{|\\ H}}{\overset{\substack{O\quad H\\ \|\quad |}}{H-O-C-C-H}}
\end{array}
$$

(written "backwards" to show the nature of the reaction), the —H of the former and O—H of the latter are pulled off, and join to form water, H—O—H. What remains of the two molecules then also join, as follows:

$$
\underset{\substack{H\;\;H\\ |\;\;|\\ H-C-C-O\\ |\;\;|\\ H\;\;H}}{}\;\;\underset{\substack{O\;\;H\\ \|\;\;|\\ -C-C-H\\ |\\ H}}{} \text{ becomes } \underset{\substack{H\;\;H\;\;\;\;O\;\;H\\ |\;\;|\;\;\;\;\|\;\;|\\ H-C-C-O-C-C-H\\ |\;\;|\;\;\;\;\;\;\;\;|\\ H\;\;H\;\;\;\;\;\;\;\;H}}{}
$$

known as the ester ethyl acetate, and used as a nail-polish remover.

The more complex esters are found in nature as fruit flavors. If the above reaction were carried out replacing the ethyl alcohol with iso-butyl alcohol,

$$
\underset{\substack{|\;\;\;|\;\;\;|\\ H\;\;H\;\;H}}{\overset{\substack{H\\ |\\ H-C-H\\ H\;\;|\;\;H\\ |\;\;|\;\;|\\ H-C-C-C-O-H}}{}}
$$

the ester produced would be

$$
\underset{\substack{|\;\;\;|\;\;\;|\\ -C-C-C-O-C-C-\\ |\;\;\;|\;\;\;|\;\;\;\;\;\;\;|}}{\overset{\substack{|\\ -C-\\ |\;\;\;\;\;\;\;\;\;\;\;O\\ |\;\;\;\;\;\;\;\;\;\;\;\|}}{}}
$$

omitting the H atoms for clarity, but again recalling that one is joined to every free dash. This is iso-butyl acetate, the flavor of raspberries. If the alcohol had been amyl alcohol,

$$-\overset{|}{\underset{|}{C}}-\overset{|}{\underset{|}{C}}-\overset{|}{\underset{|}{C}}-\overset{|}{\underset{|}{C}}-\overset{|}{\underset{|}{C}}-O-H$$

the resulting ester would have had the formula

$$-\overset{|}{\underset{|}{C}}-\overset{|}{\underset{|}{C}}-\overset{|}{\underset{|}{C}}-\overset{|}{\underset{|}{C}}-O-\overset{\overset{O}{\|}}{C}-\overset{|}{\underset{|}{C}}-$$

This compound, amyl acetate, is pear flavor. Finally, if the alcohol used had been

$$-\overset{|}{\underset{|}{C}}-\overset{\overset{\displaystyle-\overset{|}{\underset{|}{C}}-}{|}}{\underset{|}{C}}-\overset{|}{\underset{|}{C}}-\overset{|}{\underset{|}{C}}-O-H$$

iso-amyl alcohol, the product would be the ester iso-amyl acetate,

$$-\overset{|}{\underset{|}{C}}-\overset{\overset{\displaystyle-\overset{|}{\underset{|}{C}}-}{|}}{\underset{|}{C}}-\overset{|}{\underset{|}{C}}-\overset{|}{\underset{|}{C}}-O-\overset{\overset{O}{\|}}{C}-\overset{|}{\underset{|}{C}}-$$

better known as banana oil. Remember that an H atom belongs at the end of each free dash in these formulas.

SIMPLE SUGARS

Some of the sugars found in fruits can be thought of as combinations of complex alcohols, and aldehydes or ketones. **Grape sugar**, for example, has the structure

$$H-\overset{\overset{\displaystyle H}{|}}{\underset{\underset{\displaystyle H}{|}}{C}}-\overset{\overset{\displaystyle H}{|}}{\underset{\underset{\displaystyle H}{|}}{C}}-\overset{\overset{\displaystyle H}{|}}{\underset{\underset{\displaystyle H}{|}}{C}}-\overset{\overset{\displaystyle H}{|}}{\underset{\underset{\displaystyle H}{|}}{C}}-\overset{\overset{\displaystyle H}{|}}{\underset{\underset{\displaystyle H}{|}}{C}}-\overset{\overset{O}{\|}}{C}-H$$

a sort of combined penta (five O—H) alcohol and aldehyde. Fructose,

$$H-\overset{\overset{\displaystyle H}{|}}{\underset{\underset{\displaystyle H}{|}}{C}}-\overset{\overset{\displaystyle H}{|}}{\underset{\underset{\displaystyle H}{|}}{C}}-\overset{\overset{\displaystyle H}{|}}{\underset{\underset{\displaystyle H}{|}}{C}}-\overset{\overset{\displaystyle H}{|}}{\underset{\underset{\displaystyle H}{|}}{C}}-\overset{\overset{\displaystyle O}{\|}}{C}-\overset{\overset{\displaystyle H}{|}}{\underset{\underset{\displaystyle H}{|}}{C}}-H$$

found in other fruits and in honey, has the

$$\overset{\displaystyle O}{\overset{\displaystyle \|}{C}}$$

second from the end, so that it resembles a ketone, rather than an aldehyde. Indeed, grape sugar, which is also known as glucose, or dextrose, is called an aldo sugar. Fructose, because of its structural difference, is often referred to as a keto sugar.

Notice the number of asymmetric carbons contained in each molecule. If you are not awed by the complexity of these molecules, you might try modeling some of their stereoisomers on paper, for your kit does not have enough spheres to make actual models. The cane sugar we use in our coffee, known as sucrose, is even more complex. It has the formula $C_{12}H_{22}O_{11}$, which shows that most ball-and-stick kits probably do not have enough spheres to allow you to make its model.

AMINES

All classes of compounds mentioned so far are made of carbon, hydrogen, and oxygen atoms exclusively. Nitrogen, too, can bond with carbon, and give rise to still another class of compounds known as amines. They may be considered relatives of ammonia,

$$H-\overset{\overset{\displaystyle H}{|}}{N}-H$$

in which one or more H atom has been replaced by a hydrocarbon radical. The compound

$$
\begin{array}{ccc}
\text{H} & \text{H} & \\
| & | & \\
\text{H}\!-\!\text{N}\!-\!\text{C}\!-\!\text{H} \\
& | & \\
& \text{H} &
\end{array}
$$

for example, is named methyl amine. The

$$
\begin{array}{ccc}
\text{H} & \text{H} & \text{H} \\
| & \| & | \\
\text{H}\!-\!\text{C}\!-\!\text{N}\!-\!\text{C}\!-\!\text{H} \\
| & & | \\
\text{H} & & \text{H}
\end{array}
$$

molecule would be di-methyl amine, and so on. Amines are used industrially in the manufacture of drugs and dyes.

Although it is not an amine, the compound urea deserves to be mentioned here. Its structure is

$$
\begin{array}{ccc}
\text{H} & \text{O} & \text{H} \\
| & \| & | \\
\text{H}\!-\!\text{N}\!-\!\text{C}\!-\!\text{N}\!-\!\text{H}
\end{array}
$$

so that it might be described as a di-amino ketone, if such a compound were possible. Urea's importance is historical, for it was one of the first naturally occurring organic compounds to be prepared in the laboratory. Until this preparation, it had been found only in urine.

AMINO ACIDS

The last class of compounds we shall consider is unquestionably the most important to us, because they are closely related to the molecules from which we, ourselves, are made. The compounds are well named, for they are indeed both amines and organic acids. The simplest, known as glycene, has the formula

$$
\begin{array}{ccc}
\text{H} & \text{H} & \text{O} \\
| & | & \| \\
\text{H---N---C---C---O---H} \\
& | \\
& \text{H}
\end{array}
$$

Note the amino group on the left, as well as the acid group on the right. All other amino acids found in nature differ from this only in that one H of the left carbon is replaced by a group of other atoms. A second simple amino acid, for example, is alanine,

$$
\begin{array}{ccc}
\text{H} & \text{H} & \text{O} \\
| & | & \| \\
\text{H---N---C---C---O---H} \\
& \text{H---C---H} \\
& | \\
& \text{H}
\end{array}
$$

If we represent any such replacing group by the letter R, then a general formula for any of these naturally occurring amino acids would be

$$
\begin{array}{ccc}
\text{H} & \text{H} & \text{O} \\
| & | & \| \\
\text{H---N---C---C---O---H} \\
& | \\
& \text{R}
\end{array}
$$

For alanine, R is a

$$
\begin{array}{c}
\text{H} \\
| \\
\text{C---H} \\
| \\
\text{H}
\end{array}
$$

group. It is now known that proteins, one of the substances essential for life, are made of thousands of these amino acid molecules all linked together. The linking takes place, as in ester formation, by removal of an H of one molecule and an O—H of the next, to form water (H—O—H), leaving the remainder of the amino acid molecules free to link together (see Fig. 64).

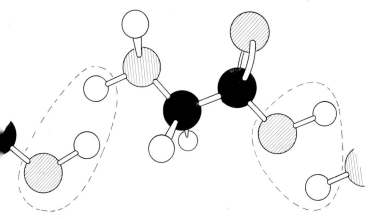

Fig. 64. Model of the simple amino acid, glycene, showing one complete molecule (center), the —O—H group of another molecule (left), and the H— of a third molecule (right). The dotted lines indicate how H— and O—H groups might be removed to form water molecules prior to the joining of the three glycene molecules into a portion of a giant long-chain protein molecule.

In the case of amino acids, we might picture the process by first lining up several amino acid molecules, then removing the necessary H and O—H atoms, and finally linking them into one long chain, thus:

This last structure is the general formula for a portion of a protein molecule. The dashes on either end of our protein indicate that there are many other amino acid links on either side of this portion before the molecule could be considered complete. As complex as such a molecule may seem, note that it is made of many simple, repeating amino acid units. One of the most exciting areas of organic chemistry today lies in the study of such molecules.

IN CONCLUSION

Of course, there are many other organic compounds which we have not even mentioned. Indeed, you now know that to mention them all would be well nigh impossible. But you do have the background which will enable you to refer to an organic chemistry text, look up for yourself the structure of any such compound and understand what you have looked up. Some compounds might be simple enough for you to model, while others will undoubtedly be much too big and complicated.

ACCURATE CRYSTAL MODELS

Introduction

This is the unit in which the great advantages of working with styrofoam spheres become most apparent. Before these uniformly accurate foamed plastic spheres became readily available, barely ten years ago, extended crystal models were made of wooden or glass spheres. This involved so much expense and difficulty that such models were found only in industry and the wealthier universities. Because of the styrofoam, a reasonably careful teenager can make a bigger, better-looking model for a dollar and several hours of his time than a geology graduate student at Harvard could build from wood in an entire term and at considerable expense.

While the detailed description of construction of the models will be given in the body of the text, it should be noted that the types of models described here are commercially available, built from wooden spheres, at a cost of some $50 per model for one of equivalent size. The fact of now being able to build one's own model for a fraction of this cost, and at the same time gain insight into the structure of the substance being modeled by the very act of modeling it, will, it is hoped, encourage the latter activity.

The plastic spheres themselves are of two types: the foamed styrofoam, or snowball, with a rough, easily compressible surface; and the smooth, hard, expanded polystyrene type. The former can be compressed, by rolling between two boards, are easier to cut, and do have the fuzzy appearance chemists like to give to

117

their conceptions of atoms. Their disadvantage is that they are often neither truly spherical nor of accurately uniform size, and that they can be deformed rather easily. The hard, polystyrene expanded type spheres are more likely to be truly spherical and uniformly accurate in size. They also show mold marks—two dots marking opposite "poles", and a circle marking their equators—which are helpful guides during their fabrication. They are more durable, cannot be compressed, and lack the fuzzy appearance already mentioned.

The question of which type of sphere to use is a matter of taste, and, to some extent, convenience. The author has found that the rough, fuzzy styrofoam is better for the molecular space-filling models described in Unit 5, because they are easier to slice (and there is plenty of slicing), and because the lack of true sphericity or exact uniformity of size is hardly noticeable in small models of this type. The harder, smoother polystyrene expanded spheres are found preferable for ball-and-stick and extended crystal models, where no slicing is required. It is also true that the accuracy of these harder spheres, both in size and sphericity, are essential to the construction of accurate extended models of crystals. Finally, the mold marks ("poles" and "equator") make fabrication much easier by providing reference marks on each sphere.

A few words now about preparation and bonding together of the spheres. Since extended crystal models each require some two dozen or more spheres, each of which are usually punched in the same manner, it will save time to do this punching completely before assembling any spheres. In order to locate a previously punched hole, it is a good idea to insert the point of a soft lead pencil in each such hole, immediately after withdrawing the punch. If you gently rotate the pencil, this will leave a dark graphite mark, easily visible on the white sphere, which will save you much time "hole hunting" during assembly.

For packing models, in which the spheres just touch their neighbors, toothpicks (the round variety), or short lengths of pipe cleaner, whose ends have been dipped in water-soluble glue, make the best connectors. The lattice-type models (like ball-and-stick) need longer connector "bonds", and short lengths of $\frac{1}{16}$-in. dowel, whose ends have been sharpened and dipped in glue, prove satisfactory. A ready source of such dowels are the drugstore medical applicators, previously mentioned.

Because of the large number of spheres required for each crystal model, and because each type of structure described on the following pages is representative of many substances in addition to the one being modeled, it makes sense to think of each such model as representing any or all of the substances showing its structure. The sodium chloride, or rock salt arrangement, for example, which is the first ionic structure considered in this unit, is shown by over 150 different chemical compounds.

If one views each model in this light, an exact scale is unnecessary, and can be subordinated to the sphere sizes most readily available. As an example, since the chloride ion is thought to be almost twice the diameter of the sodium ion, and this ratio is approximated in many other sodium chloride type structures, spheres representing chloride ions need be only about twice the diameter of those representing sodium ions. They do not need to conform, necessarily, to a particular scale. If, on the other hand, you wish to have an accurate scale model, of, say, sodium chloride, you need only look up the accepted ionic radii in the Appendix, and choose your sphere sizes accordingly.

Painting of spheres, a formidable task for a large number of spheres, can also be simplified. If only one type of atom makes up a substance, the spheres modeling it could well be left unpainted. For ionic crystal models, the different kinds of ions could be distinguished by painting only, say, the spheres representing the positive ions. Thus the colors would not limit the model to representing only one substance, and both time and paint would be saved. Again, for those readers who wish it, each model could represent only the one substance, and be scaled and painted accordingly. As already mentioned, at least a first coat of paint must be water-soluble, latex based. A second coat could be enamel, providing the first coat completely covers the sphere.

120 *Atomic and Molecular Models*

When faced with the task of painting a large number of spheres the same color, the model-maker is advised to adopt the following procedure. First, fabricate (i.e. punch all necessary holes in) each sphere. Then insert toothpicks into each punched hole. Pour paint into a large, deep pan, and—using toothpicks as handles, —dip each sphere into the paint, completely submerging it. Remove dipped sphere, let drip for several seconds over the pan, and then rest it on its toothpicks on a piece of wood, or a piece of sheet styrofoam, available from the same sources as the spheres themselves. This is much quicker than hand-painting each sphere, and the toothpicks will preserve the hole marks you have made, allowing them to be located with ease when needed. When dry, the spheres can be assembled into the finished model.

This unit assumes only an elementary background in chemistry. Concepts such as density, gram-atomic and gram-molecular weight, the mole, the concept of positive and negative ions, as well as elementary geometry, are used here without attempt at development. The first part of the unit describes detailed construction of models representative of typical crystal structures. This is followed by some examples of the pedagogic possibility of these models.

Although already mentioned in Unit 5 (pp. 41–44), the reader is again reminded of the difference between lattice and packing models. The lattice model is simply a ball-and-stick model of a portion of a crystal. As such, it shows the location of the centers of the atoms or ions, as well as the manner in which they are bonded together. The packing model is analogous to the space-filling type, where the so-called effective size of the atoms or ions is to scale. In such models, the adjacent spheres touch each other, giving perhaps a more accurate idea of the atomic dimensions, but hiding the interior portions of the model.

We have already discussed van der Waals and covalent radii but this unit introduces two other types: metallic and ionic Metallic radii are analogous to van der Waals radii because they are simply half the internuclear distance between neighboring atoms in a metallic crystal. Ionic radii can also be thought of as van der Waals—like in that they define the effective sizes of ions. Their determination is, however, less direct, and will be considered in some detail later on. The nice thing for the model-maker is that in both metallic and ionic crystals, the atoms and ions are thought to be in contact, but not to overlap each other. Practically, if we assume a sphere represents the effective size of such an atom or ion, this means no slicing off of portions of the spheres, even for the packing models.

THE UNIT CELL CONCEPT

All substances whose structures are represented by the models described in this unit are crystalline. They have sharp melting points and other properties which are best interpreted by imagining them to consist of regular arrays of atoms or ions. They can thus be thought of as three dimensional analogues of large sections of wallpaper, since both are made up of a basic pattern, repeated over and over again. Since it is obviously impossible to talk of or model anything more than a small portion of such a structure, chemists find it useful to define such a small portion, which they call the unit cell.

The unit cell is analogous to the basic pattern of a wallpaper in the sense that both the crystal and the wallpaper can be viewed as being made of numerous unit cell patterns, stacked together in a regular way. It makes sense, therefore, when modeling the structure of a particular crystalline material, to make the model at least one unit cell in size. This convention will be followed in a majority of the models considered here.

GENERAL METHOD OF MODEL CONSTRUCTION

It seems obvious that, for each substance to be modeled, one must consult the literature to determine the internal structure of the crystal, as well as atomic and unit cell dimensions. From this information, the model-maker can select a suitable scale, and hence decide on the sizes of the spheres he will use. Finally, the spheres are fabricated so as to show the correct structure when assembled. This, of course, can always be done. Although some structures require a good deal of visualization in the transfer from two-dimensional page to three-dimensional model, the process is both interesting and rewarding—the best way to really come to know that structure.

As already indicated in the introduction to this unit, it turns out that many common crystalline substances show one of several basic structures. The examples which follow are representative of the better known of these basic structures. Once the interested reader has had the experience of modeling these more basic structures, he is encouraged to try on his own some of the more exotic arrangements he will find in the literature.

CONSTRUCTION AND USE OF JIGS

The basic structures we shall consider are geometrically simple because of the ways in which the atoms or ions in them are surrounded (coordinated) by their neighbors. In the model, this means a given sphere will be joined to neighboring spheres in one of only several simple ways. If we consider the geometric centers of the given, and surrounding spheres, it turns out that the former is at the center, and the latter, at corners, of such common geometric solids as a cube, tetrahedron, or octahedron. Or, considering one layer of spheres, their centers lie in one plane, and the central sphere's center is surrounded by spheres whose centers lie at corners of a hexagon or square.

It is hardly surprising, therefore, that—since each sphere in a model can be considered a central one surrounded by neighbor spheres—the job of preparing a sphere so it can be joined to its neighbors boils down to one of several possibilities. We have already considered the problem of punching holes in a sphere so its toothpick or dowel "bonds" would extend from it in the proper directions to result in an accurate, finished model. Recall, for example, the preparation of spheres representing oxygen, nitrogen, and carbon, as detailed on pp. 49–57, and Figs. 22 and 24. Note the use of jigs, described on pp. 49 and 57, and Fig. 22, and again on p. 63, Fig. 32.

The jigs used in those constructions are identical to the ones you will need for the crystal models described in this unit, except for the radii of the cut-out circles. To make them, obtain seven approximately 5 × 7 in. pieces of heavy, carton-type cardboard. Using protractor and ruler, carefully pencil on three pieces the lines shown in part (a) of Fig. 65. Pencil part (b) lines (Fig. 65)

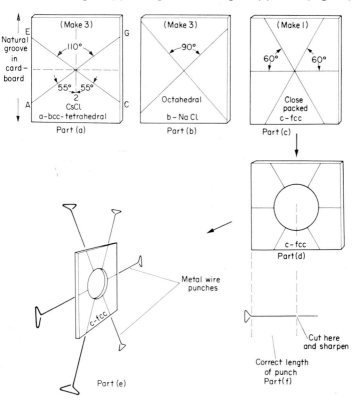

FIG. 65. Jig patterns to be traced onto packing-carton-type cardboard. Eventually, circles whose radii will match spheres to be used, will be cut out (part d), and the jig used as shown, with wire punches, in part (e).

on three other pieces, and place part (c) lines on the remaining cardboard. Make sure that no solid line runs parallel to the natural grooves between the outer layers of the cardboard, so that when metal wire (welding rod) "punches" are inserted, the cardboard will hold them firmly.

Eventually, a circle, whose center is at the intersection of the solid lines in the jig patterns shown in Fig. 65, will be drawn with a compass on each cardboard, and cut out with a razor to give (for part (c)) the appearance shown in Fig. 65 part (d). Since the radius of each cut-out circle must be identical in size with the radius of the sphere to be inserted in it, do not draw or cut out these circles until you are sure of the sphere sizes you are going to use. When this is known, the circles may be prepared and cut, and metal punches inserted along the direction of each solid line, but between the layers of cardboard so that the cardboard itself holds the punches in place. The punches are best made from $\frac{1}{32}$-in. diameter welding rod, or an equivalent sized, stiff metal wire, cut into small lengths, and with one end bent into a triangular handle with pliers (Fig. 65, part (e)). To make these punches the proper length, lie each on the jig along the line it will be inserted along, so that its handle lies at the jig's edge (Fig. 65, part (f)). Snip the wire at the center of the cut-out circle, and sharpen it with emery paper or a knife sharpener. When inserted in the jig, each punch should just reach the geometric center of the jig when pushed all the way in, and all such punches should meet at this geometric center.

MODELING CLOSE-PACKED METALS

Most metallic elements crystallize in one of two almost identical so-called "close-packed" arrangements. Here, as the name implies, the atoms are as crowded as possible, and are arranged in layers (Fig. 66) so that the atoms of one layer rest in the depressions between the atoms of the one underneath it. Notice that third layer atoms can rest either directly over first layer atoms, or over the holes between them—hence the two possibilities.

To model such arrangements, one could simply join three pieces of cardboard (Fig. 67 part (a)), pack a layer of spheres within, and proceed to stack additional layers of spheres above it (Fig. 67, part (b)). You have no doubt noted that each sphere in

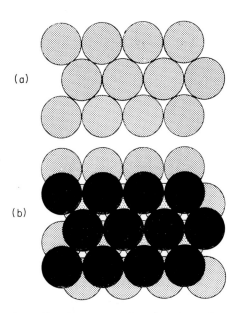

(a)

(b)

FIG. 66. (a) Looking down at a portion of one layer of close-packed metal atoms. (b) Positions of a second layer of close-packed atoms (dark circles) when added to a first layer. Note that third layer atoms could be placed either directly over first layer atoms, or over the holes between them. Cubic close packing is the result of this latter arrangement.

such an arrangement is surrounded in its own layer by six neighbors (Fig. 67, part (c)) whose centers lie at the corners of a regular hexagon. If a more convenient model than the one shown in Fig. 67, part (b) is desired, it can be made from layers (Fig. 68, part (a)) in which each sphere has been punched with the jig shown in parts (c), (d) and (e) of Fig. 65. Toothpicks, or short

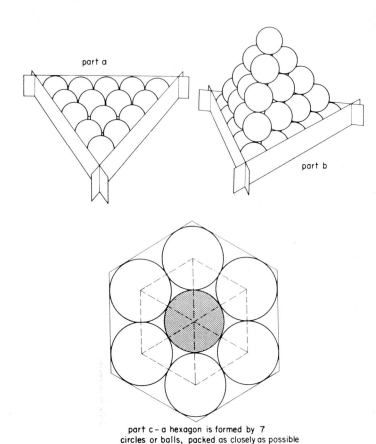

part a

part b

part c – a hexagon is formed by 7
circles or balls, packed as closely as possible

FIG. 67. Close packing of styrofoam spheres. A quick but imper-
manent model is made by joining three lengths of manila paper as
shown, and then stacking spheres within, as shown. Part (c) shows
the coordination (surroundings) of each of the spheres by neighbors
in its own layer.

lengths of pipe-cleaner, are then inserted in the punched holes, and joined to give the layer shown in Fig. 68, part (a), where the dotted lines represent the "bonds". If joining the spheres by using all the bonds (Fig. 68a) proves too awkward, an alternate procedure is suggested in Fig. 69. This gives a layer of spheres

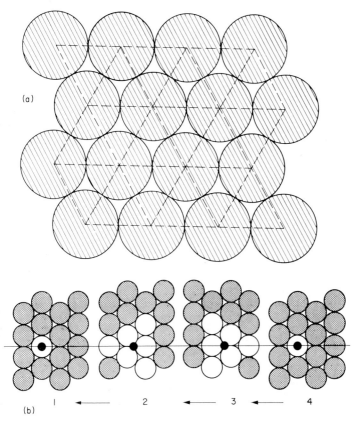

FIG. 68. (a) Suggested layer arrangement for close-packed model. Four such layers should be made. If placed above each other as shown at (b), the cubic unit cell would be represented by the white spheres, providing the imaginary black dots lay directly above each other.

almost as strongly held together, and considerably easier to assemble. Each sphere is punched the same way as above, but some of the punched holes are left empty of "bonds".

If four such layers are made, they may be stacked on top of one another so that layer 3 goes directly over layer 1, giving an arrangement called hexagonal close-packing. Here, since layer 3

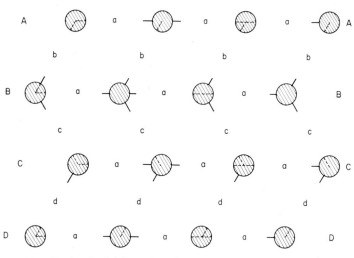

FIG. 69. Plan for joining spheres into layer shown in Fig. 68. Insert toothpick, or pipe-cleaner bonds, as shown by short solid lines. Then join spheres along the "*a*" directions into four lines of spheres, lines *A*, *B*, *C* and *D*. Then join the lines *A* and *B* along the "*b*" directions, and finally add to them line *C* along the "*c*" directions, and line *D* along the "*d*" directions. Dotted lines show where bonds will penetrate when layer is completely assembled.

is a "repeat" of layer 1, and layer 4 a "repeat" of layer 2, the notation 1–2–1–2–1–2, etc., indicates this basic repetition of layers in an imagined extended stacking of layer upon layer. Its name is derived from the hexagonal symmetry (Fig. 67c) within each layer.

The model described above may be rearranged, since one layer is not permanently attached to another, so that layer 3 spheres are directly over the holes between layer 1 spheres. When this is done, layer 3 is NOT a repeat of layer 1, for it is layer 4 spheres which now lie over layer 1 spheres. Thus the pattern has become 1–2–3–1–2–3–1–2–3, etc., an arrangement known as cubic close-packing. This name comes from another, cubic, pattern now present in the model, and to visualize it imagine that the four layers you just assembled were made in part of different color spheres, such as the white and dark ones shown in Fig. 68, part (b). When such layers are stacked so that the black spots of Fig. 68b are vertically above each other, the fourteen light spheres shown there form a cube. As it sits in those four close-packed layers, it is tilted over, but still of course a cube. It is, in fact,

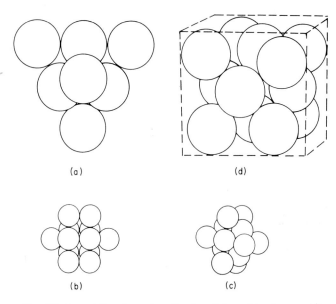

FIG. 70. Assembling the unit cell cube of the cubic close packing arrangement, as a separate model. (a) Make two identical, seven-sphere units, each as shown. Then (b) hold them together, base to base, and rotate one form (c) until its base spheres nest between those of the other. Turn form as shown (d) to note its cubical shape.

the so-called unit cell of this type of structure. You can assemble this unit cell cube by assembling two 7-sphere units, each as shown in Fig. 70a; then rotate them, back to back (Fig. 70b and c) until they fit together to form the cube shown in Fig. 70d.

You may have noticed that no scale has been suggested for the models described above, and this is in line with the points made on p. 119. Even if you choose to keep the four layers permanently unattached, so they may be used to model both hexagonal and cubic close-packing, they call for sixty-four spheres, not to mention the fourteen needed if the unit cell cube is to be modeled. In view of the higher cost of larger spheres, it is suggested that approximately $1\frac{1}{2}$-in. (4 cm) diameter spheres be used for these models. It is also suggested that only the packing models of these structures be attempted. Not only would a lattice model require a large number of additional spheres, but the three-dimensional bonding required would prove extremely difficult. The Appendix lists those metals believed to crystallize in the hexagonal, as well as those having the cubic close packing arrangements.

THE SODIUM CHLORIDE STRUCTURE

Known also as the rock salt arrangement, this is probably the most well known of all crystal structures. Some 150 different chemical compounds crystallize in this way, and a few of these compounds are listed in the Appendix. The evidence suggests that the units in these structures are ions—atoms that have lost or gained electrons to become positively or negatively charged particles. Although we pick the commonest example of this arrangement, ordinary table salt—sodium chloride (NaCl)—to model, remember that your model can also represent any one of the other substances that show its arrangement.

Since sodium chloride, NaCl, is made of positive and negative ions, it is hardly surprising—in view of the laws of electrostatic attraction—to find each positive sodium ion in the arrangement surrounded by negative chloride ions, while each chloride ion is similarly surrounded by sodium ions. The number of oppositely charged ions surrounding a given ion in sodium chloride is six, and this should remind you of your Unit 5 work with models of complex ions. Refer back to p. 63 and Figs. 32–34 for examples of how a central sphere is surrounded symmetrically by six neighbors, and note that these neighbors' centers are at the corners— the central sphere at the center—of a regular octahedron (Fig. 33). This is just the arrangement, or coordination, in sodium chloride, as shown in Fig. 72b, and means that all spheres must be prepared in the same manner as was described on p. 63 and Fig. 32.

Both the packing and lattice models of sodium chloride shown in Fig. 72 require identical preparation of their spheres—all on the NaCl jig (part (b) of Fig. 65), which is of course identical to the octahedral jig of Fig. 30b. It is suggested that $1\frac{1}{4}$-in. (3.2 cm) diameter spheres, painted a dark color, be used for sodium ions in both packing and lattice models (you will need twenty-six). For chlorides, a $2\frac{3}{8}$-in. (6 cm) diameter sphere is recommended for the packing, and a $1\frac{1}{2}$-in. (3.8 cm) diameter sphere for the lattice model. You will need fourteen of each size, and these can be left white. If you are troubled because neither the sizes nor suggested colors are consistent with those used in earlier models, recall that the sodium chloride structures are typical of many compounds, any one of which can be represented by your models. We have kept in the packing model, the approximate one-to-two sizes of the sodium-to-chloride ions. Those who wish an accurately scaled model of sodium chloride only need turn to the ionic radii given in the appendix, and choose their sphere sizes accordingly.

Once you have decided upon the sizes of your spheres, you can draw and cut out the proper-sized circles on your three NaCl jigs (Fig. 65b): one for the sodium, Na^+, ions; one for the packing chloride, Cl^-, ions; one for the smaller lattice chloride ions. All spheres are punched just as described on p. 63, and shown in Fig. 32: first with all four punches, then withdrawing any two opposite punches, rotating the sphere 90° about the other two, and then reinserting the two withdrawn punches to make holes 5 and 6.

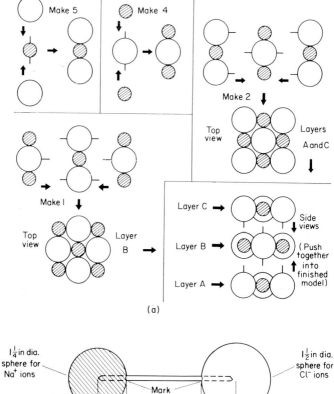

(a)

(b) Suggested dowel-rod detail for lattice model of NaCl

Fig. 71. NaCl model assembly detail. (a) Assemble packing model by first joining lines of spheres (top left), then pushing lines into layers (center left and top right), and finally pushing layers together (lower right). (b) Shows dowel detail and suggested sphere sizes for lattice model, which is assembled in the same order as the packing model.

FIG. 72. Packing (left) and lattice models (right) of sodium chloride unit cell. Note that central dark sphere of latter, representing sodium ion, has six nearest neighbors (white chloride ions) arranged octahedrally about it, as shown in center of photo.

Do this as carefully as you can, using the mold marks on the spheres to help guide you, and remember to use a soft lead pencil to mark each hole you have made.

Suggested assembly procedures are shown in Fig. 71. Before this, however, it would be wise to insert toothpicks into the punched holes of the "sodium ions" and dip all twenty-six of these into some dark, water-based paint: red, blue or even an aluminum-type paint. After they have thoroughly dried, remove the toothpicks and permanently insert others as shown in Fig. 71a, first dabbing the end to be inserted with water-soluble glue. Be sure the toothpicks do not go more than half-way through the sphere (i.e. to its center), and that the part left sticking out is not longer than the radius of the sphere it will eventually go into (this would be $\frac{3}{4}$-in., or 1.9 cm for the suggested scale), by cutting off their ends if necessary. Assemble the layers as shown, and finally press layers together, being sure toothpicks fit previously punched holes.

The lattice model is prepared just the same way as the packing model was, except for the smaller "chloride ion" spheres, and the use of $\frac{1}{16}$-in. dowel rods in place of toothpicks. A dowel length is suggested in Fig. 71b. Cut fifty-four pieces of dowel, each 7.5 cm in length, and mark each with a soft pencil, 1.6 cm from one end, and 1.9 cm from the other. Sharpen each end, and following the procedure outlined for toothpicks in Fig. 71a, insert dowel into proper sphere, but only up to the mark you have made. A dab of water-soluble glue placed on the sharpened end just before insertion will insure a securely held-together model. The finished model should resemble that in the photo of Fig. 72c.

These models, of course, represent only a single unit cell of this so-called sodium chloride structure. If you have the spheres and the patience, you may wish to build models that include more than the unit cell. Such packing models are shown in Figs. 73 and 74. But regardless of where practical considerations force you to stop the model, each ion in the actual arrangement is surrounded by six others of opposite charge, whose centers are at the corners of a regular tetrahedron (Fig. 75).

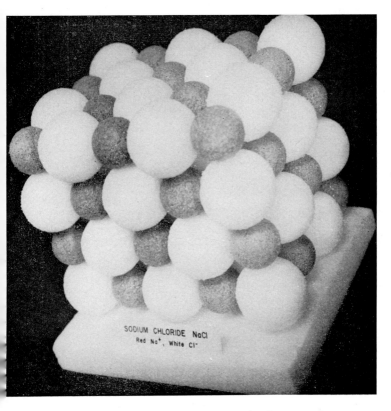

FIG. 73. Packing model of NaCl structure, showing more than a single unit cell.

Fig. 74. Extended packing model of the sodium chloride structure.

FIG. 75. Model showing that central "chloride ion" at right is considered octahedrally bonded to its neighbors because bond directions extend (center) to the vertices of an octahedron like one pictured at left.

FIG. 76. Model of the unit cell (right) of the CsCl arrangement, showing why it is an example of the so-called cubic body centered type. Note that each Cs$^+$ ion (dark sphere) is bonded to eight Cl$^-$ ions (middle) whose geometric centers, when connected, form a cube (left). The Cs$^+$ ion sits in the center of the body of this cube, and hence the name.

CUBIC BODY CENTERED BONDING

Li, Na, K, Rb, Cs, CsCl

THE CESIUM CHLORIDE ARRANGEMENT

Though most of the alkali halides show the sodium chloride arrangement, some do not. Cesium chloride, for example, crystallizes at ordinary temperatures so that each cesium is surrounded by eight chlorides, and each chloride is surrounded (coordinated) by eight cesium ions. Geometrically, one may think of the cesium ion at the center, and the eight chloride ions at the eight corners, of a cube (Fig. 76). The smallest models that show this are shown in Fig. 78 center (packing) and right (lattice). Since most substances showing this structure have positive and negative ions of approximately the same size, or are made of identical atoms (the alkali metals show this arrangement), all spheres chosen should be of the same diameter. To be consistent with the sodium chloride model, with its chloride ions represented by 6-cm (packing) and 3.8-cm (lattice) diameter spheres, we might choose these same size spheres (to represent both cesium and chloride ions) for our cesium chloride models.

Each ion in this arrangement may be pictured as at the geometric center of a cube (at "O" in Fig. 77), surrounded by oppositely charged neighbors at the eight corners (A–H in Fig. 77). The dotted lines in the figure, the so-called "bonds" between ions, lie in two of the cube's diagonal planes ($AEGC$ and $BFHD$ in Fig. 77), which intersect each other at right angles. If one looks at the jig pattern shown in Fig. 65a, it is seen that its two intersecting lines represent directions of the four "bonds" in one of these planes. This jig, after cutting out a 3-cm radius circle on one and a 1.9-cm radius circle on a second cardboard with this pattern, may be used to fabricate spheres as follows: center the sphere in the cut-out circle, and punch along all four lines (A, E, C, G in Fig. 65a). Withdraw all punches, and carefully rotate the sphere 90° about its vertical (1–2 in Fig. 65a) axis, until the holes you just made lie in a plane at right angles to the plane of the jig. Then punch the sphere again along all four directions. The eight holes thus made will extend from the sphere's center to the eight corners of an imaginary, circumscribed cube.

The identical procedure is used to prepare both lattice and packing model spheres. Mark each hole with a soft pencil, and plug the holes of eight of the small (lattice) and eight of the large (packing) spheres so prepared with toothpicks. These will represent the cesium ions, and must be dipped in a dark, water-based

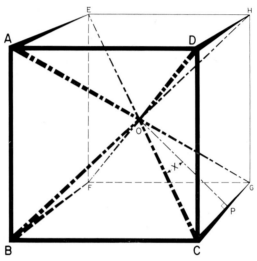

FIG. 77. Cube illustrating structural unit of cubic body centered arrangement. All bonds would lie in two of the cubes diagonal planes, which intersect at right angles. Cube also shows how the jig (Fig. 65a) required for construction of models of this arrangement is prepared. Solution of triangle *OPC* will yield jig angle *X*. The letters used here correspond to those used in Fig. 65a.

paint, and then allowed to dry. The eight other large and eight other small spheres you should have prepared (sixteen for each model) will represent the chloride ions, and may be left an unpainted white.

Assembling the models can be simplified in several ways. First, it is suggested that the packing model be assembled partially, using short lengths of pipe-cleaner for bonds, rather than toothpicks, because their greater flexibility will make the work easier.

Fig. 78. Packing (left and center) and lattice models of the body-centered cubic cesium chloride structure. The unit cell is shown at left, but model including two possible, overlapping, unit cells (center) gives a better idea of the extended arrangement.

Cut twenty-four 6-cm lengths of pipe-cleaner for this purpose. Second, it is a good idea to lightly pencil numbers on each sphere of each type model, as shown in Fig. 79. The packing model assembly is identical to that of the lattice model, shown in the figure, so that the same set of numbers can be used for both. They will aid you in locating and joining the spheres, and can be erased after assembly. Place pipe-cleaner lengths in the spheres' pre-

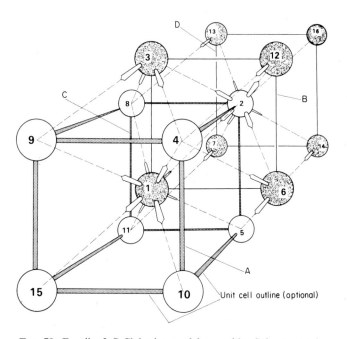

FIG. 79. Detail of CsCl lattice model assembly. Spheres can be pictured as in one of four vertical planes, each parallel to the plane of this paper: spheres 4, 9, 10, 15 in the foremost; spheres 1, 3, 6, 12 in the next; spheres 2, 5, 8, 11 in the next; and spheres 7, 13, 14, 16 in the rearmost such plane.

Spheres 1 and 2 each make eight bonds with neighbors (one of each of these bonds is to each other), spheres 3–8 each make four bonds, spheres 9–14 each make two, and spheres 15 and 16 make only one bond apiece. Shaded lines represent imaginary unit cell boundaries—NOT actual bonds. Their use is optional.

pared holes as shown in Fig. 79, and begin to join the spheres
as shown. Do not use any glue as yet, and do not push the spheres
completely together. Thus unfinished, this model should help you
to prepare the lattice model.

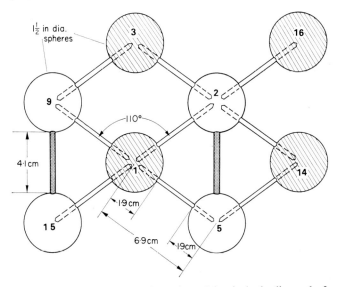

FIG. 80. Cross-section showing spheres lying in body diagonal of
unit cell cube illustrated in Fig. 79. Sphere numbers correspond to
those in the previous figure. Dowel detail is shown, but includes only
dowels lying in the plane being pictured. Sphere 1, for example, has
four more dowels in it, in a pattern identical to those shown, but at
an angle of 90° to them. In other words, these four "invisible" dowels
would come directly out of, and behind, the plane of this paper.
Shaded dowels are the unit cell outlines that would lie in the plane
of the spheres being shown here.

For the lattice model, whose dowel detail is shown in Fig. 80,
number with a pencil the smaller spheres, just as you did the
larger ones. Cut twenty-four 6.9-cm length dowels, marking
them 1.9 cm from each end, and then sharpening the ends.
Partially insert them into each numbered sphere as shown in
Fig. 79, and carefully assemble the spheres as shown in the

figure. Be patient at this stage, and use no glue until you are sure
of the assembly. Then, and only then, place a dab of glue where
dowel enters sphere, and push dowel in up to the mark on it.
The packing model may now be finished by placing a dab of
glue where each pipe-cleaner enters a sphere, and then pushing
the spheres completely together to resemble the model shown at
center of Fig. 78.

If you wish to show the unit cell outline (shaded dowels in Figs.
79–81), it is suggested that you cut twelve 4.1-cm dowel lengths,
paint them in some way to distinguish them from the dowel bonds
previously used, and insert them between the white spheres as
shown in Fig. 79. Dab their unsharpened ends with glue, and

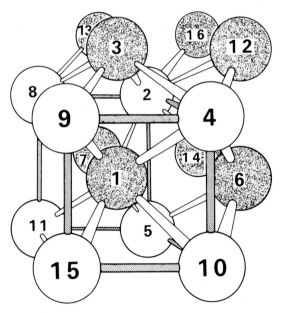

FIG. 81. Diagram of the finished lattice model, in the same per-
spective as the model photographed in Fig. 78, right. Sphere
numbers correspond to those used in Figs. 79 and 80, and the
shaded dowels again represent unit cell outlines.

hold in place until dry. Do NOT attempt to push these unsharpened dowels into the white spheres. These unit cell outlines
are of course fictional, and arbitrary. They might just as well
have connected the eight dark spheres. They are not to be con-
fused with actual bonds, and their use here is strictly a matter of
taste. The photo of the lattice model at right of Fig. 78 does have
these unit cell outlines, as does the drawing of the finished lattice
model, Fig. 81.

An approximation to the cesium chloride structure is shown in
Fig. 82. It is suggested for readers wishing to make extended

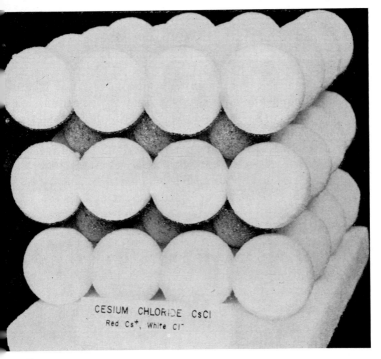

Fig. 82. An approximation to the cesium chloride structure, made
by placing dark (red) spheres in depressions between octahedrally
punched layers of white spheres. A scale model would show even
spacing between all white spheres.

packing models of this arrangement, because of its relative ease of construction. To model it, punch only forty-eight 6-cm diameter white spheres on the sodium chloride jig (Fig. 65b) exactly as you did the chloride ions in sodium chloride, except that you need not rotate the sphere, nor punch a fifth and sixth hole. Simply make the original four punches. Then join these spheres into layers (the white spheres in Fig. 82). Place four dark spheres, slightly smaller in diameter (5.1 cm) than the white ones, in the depressions between white spheres, as shown in Fig. 82. If desired, these layers may be glued together where the spheres contact each other.

CALCIUM FLUORIDE AS AN EXAMPLE OF THE FLUORITE ARRANGEMENT

Calcium fluoride, CaF_2, is an example of the so-called fluorite structure, adopted by some twenty compounds with the general formula MX_2. It is related to the previously discussed cesium chloride arrangement in that both the calcium- and fluoride-represented spheres must be prepared using the same type jig. This is due to the coordination of each type sphere, and the relation between tetrahedron and cube. In calcium fluoride, each calcium ion is coordinated (surrounded) by eight fluoride ions—in just the way each cesium was surrounded by eight chlorides in the cesium chloride structure. Each fluoride ion is, in turn, bonded to four calcium ions such that if the fluoride were at the center, the calcium's would be at the corners of a regular tetrahedron (Fig. 25).

The reason the same jig (Fig. 65a) can be used to prepare spheres representing both Ca^{++} and F^- ions becomes apparent when one examines Fig. 26(a) and 83. Note that when a tetrahedron (shaded lines in Fig. 83) is inscribed in a cube (solid lines in Fig. 83), they share the common geometric center, O. Note also, that when this center, O, is joined to the four corners of the tetrahedron (A, C, F, and H in Fig. 83), this represents only every other line joining center O with the eight corners of the cube (see Fig. 77). Thus, to fabricate spheres representing Ca^{++} ions, one would use the four punches of jig (Fig. 65a) twice, as described for the cesium chloride models, making the required eight holes. For F^- ions, however, every other punch would be used, making four holes in all. This, you will recall, is the way NH_3 and CH_4 were prepared, as described in Figs. 21(b) and 23.

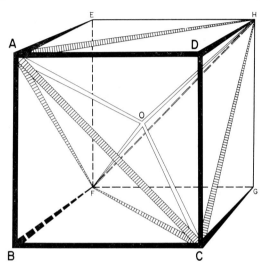

FIG. 83. Tetrahedron (shaded) inscribed in cube (black), showing that they share the same geometric center, O, and four of the eight verticies of the cube.

Again, both packing and lattice models of CaF_2 are prepared in the same manner, except for the smaller sphere sizes in the lattice model. Recommended sphere sizes are: calcium's 1.9-cm radius (packing) and 1.6-cm radius (lattice); fluorides: 2.55-cm radius (packing) and 1.9-cm radius (lattice). This scale is almost the same as was used for the NaCl and CsCl packing models, is close (in the packing model) to the same ratio as are the ionic radii of 0.99×10^{-8} cm for Ca^{++} and 1.36×10^{-8} cm for F^-. Also, it will allow you to use one of the jigs already prepared for

148   *Atomic and Molecular Models*

the CsCl lattice model. You will need to prepare two other jigs with the pattern shown in Fig. 65a, but with 1.6 and 2.55-cm radius cut-out circles, if you use these suggested sphere sizes. It is recommended that spheres representing Ca^{++} ions be dipped in a dark, water-based paint, while the F$^-$ spheres be left unpainted.

The spheres are prepared as follows: select fourteen 1.9 cm radius spheres and fourteen 1.6 cm radius spheres to represent calcium ions. Place sphere in proper-sized jig and punch ONLY

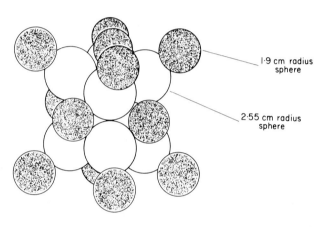

1·9 cm radius sphere

2·55 cm radius sphere

FIG. 84. Drawing of packing model of calcium fluoride, CaF$_2$, the so-called "fluorite" arrangement, according to Wyckoff. Suggested sphere radii are 1.9 cm for the Ca^{++}-representing dark spheres, and 2.55 cm for the F$^-$-representing white ones.

with the "*A*" and "*E*" punches (Fig. 65a), rotate sphere 90° about the "1–2" axis, and again punch only with "*A*" and "*E*" punches. The reason for omitting the other four ("*C*" and "*G*") punches is purely the artificial one of limiting our model to one unit cell in size. Such a unit is outlined by the dotted lines in the model at lower left of Fig. 85. Note that the dark spheres (representing Ca^{++} ions) in the center of every face of the (dotted lined) outlined cube is joined to four white spheres (F$^-$ ions) contained in the body of this cube. In an extended model, they would be joined

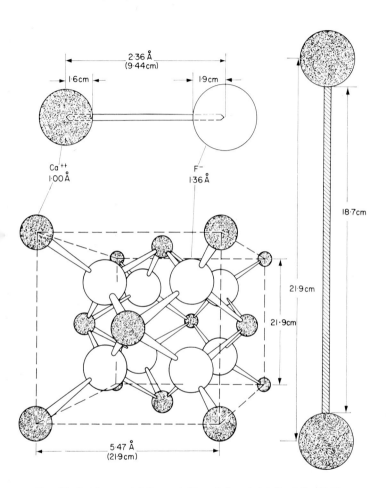

Fig. 85. Lattice model (lower left) and dowel detail of the CaF_2 unit cell. Dark spheres represent Ca^{++} ions, and dotted lines in model (lower left) show unit cell outline.

as well to four white spheres in adjacent unit cells. Since these adjacent units are not represented in our model, we may omit the four other "bonds". Actually, since there are only six such face-centered spheres for each model, only six of each sized Ca^{++} representing sphere need be punched in this way. The other eight spheres for each model need only be punched once, with any of the jig's punches.

To prepare the larger, white, F^--representing spheres, select eight 2.55-cm and eight 1.9-cm radius spheres. Punch each of these spheres, on the proper-sized jig, by centering sphere, and punching along the "*E*" and "*G*" lines shown in Fig. 65a. Then rotate the sphere 90° about its "1–2" axis, as shown in the figure, and finish punching with the "*A*" and "*C*" line punches. This makes the four required tetrahedral holes (see Fig. 83). These latter spheres are best left unpainted. The former, Ca^{++}-representing spheres should have their holes plugged with toothpicks, then be dipped in a dark, water-soluble paint, and allowed to dry.

For final assembly of the models, cut thirty-two 4.4 cm lengths of pipe-cleaner, and thirty-two 9.44 cm lengths of dowel. Mark 1.9 cm from each end of the latter, and sharpen these ends. Assemble both packing and lattice models as shown in Fig. 85, using pipe-cleaner for the packing model bonds, but pushing neither pipe-cleaner nor dowel rods as far into the spheres as they will finally go. When the models resemble the drawings in Figs. 84 and 85, place a dab of glue where, in turn, each dowel, and each pipe-cleaner, enter the sphere, and push dowels in up to their marks, pipe-cleaners in until spheres touch. If you wish to place the imaginary unit cell outline on your lattice model, cut twelve 18.7-cm lengths of dowel, do NOT sharpen the ends, but do paint them so as to distinguish them from the other, bond-representing dowel rods. Then place them between corner spheres in the lattice model, as shown at right of Fig. 85. A dab of glue between end of dowel and sphere will serve to hold these unit-cell representing dowels in place. Care must be taken not to distort the corner spheres when placing these long dowels between them. Again, the use of this unit cell outline is a matter of taste, although it does clearly show why it is named a face-centered cubic unit cell. The dark sphere in the center of each such cube face is, of course, responsible.

An alternate, but far-easier-to prepare packing model of CaF_2 may be prepared by punching twenty-seven of the 2.55-cm radius white spheres on an octahedral, NaCl-type jig (Fig. 65b) with the proper-sized cut-out circle. This will require making another Fig. 65b-type jig, punching each sphere with all four punches, then withdrawing any two opposite punches, rotating the sphere 90° about the other two punches, and reinserting the two previously withdrawn punches. Assemble the spheres into three 9-sphere layers, as shown in Fig. 86, placing four 1.75-cm radius dark spheres on the bottom and middle white-sphere layers as shown in Fig. 87. Toothpicks placed in each of the two vertical holes of the four corner spheres of the middle layer (the X marks in Fig. 87) will then hold the layers together when they are pushed into place. This model does show the eight white spheres surrounding each dark sphere, and, for the white sphere in the model's body center (sphere "*O*" in Fig. 87), it does show its four surrounding dark spheres. The dark spheres' slightly smaller size is simply to allow them to sit in the holes between white spheres.

LATTICE MODEL OF GRAPHITE STRUCTURE

The familiar graphite form of carbon is thought to be made of layers of carbon atoms, bonded strongly to one another within a layer, but with only weak bonds between layers. This nicely explains the slippery feel of graphite as due to the easy sliding of one layer over another. To model it, select fifty-seven 1.3-cm radius spheres, which may be left an unpainted white. Prepare a cardboard jig with the pattern shown in Fig. 30(a), but with a 1.3-cm radius cut-out circle, and punch twenty-four of the spheres with all three punches, the remaining thirty-three spheres with any two of the punches. Mark eight of the 3-hole-punched spheres, and four of the 2-hole-punched spheres with a soft

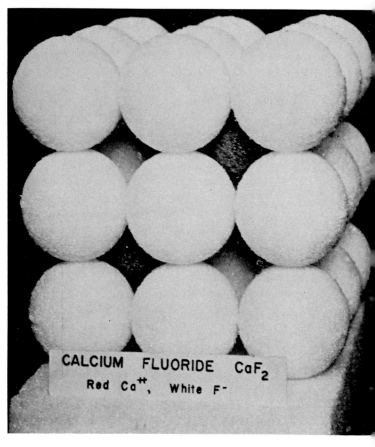

Fɪɢ. 86. Photograph of an alternate, but equally valid packing model of CaF_2, according to Pauling, and using the white F^--representing spheres to outline the unit cell. Dark spheres, representing the Ca^{++} ions, are placed in alternate holes between white spheres (see Fig. 87).

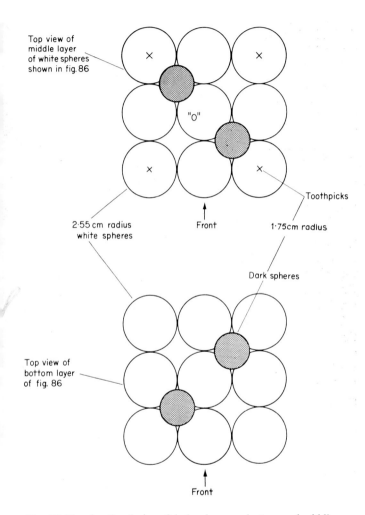

FIG. 87. Showing the placing of dark spheres on bottom and middle layers of white spheres shown in Fig. 86. View is from above.

pencil, and punch holes in each completely through the sphere's center, but at right angles to the plane of the previously punched holes. Do this slowly, rotating sphere to check for "wobble" as shown in Fig. 11. These twelve marked spheres will correspond to the darkened ones shown in Fig. 88.

Cut sixty-nine 5.68 cm lengths of dowel, mark each 1.3 cm from each end, and then sharpen the ends. Use these dowel rods to assemble the spheres into three separate, identical, layers, as shown in Fig. 88. Note that the middle layer, layer 2, is the same as the others, but has been turned from left to right, and then over. Be sure that the 3-punched-holes spheres are used only where three dowels are required, and that the marked spheres correspond in position to the darkened ones of Fig. 88.

This is one model where the unit cell outline, dotted in Fig. 88, is needed to hold the layers together. To simulate it, cut four 26.2-cm lengths, and eight 7.2-cm lengths, of dowel, painted so as to distinguish them from the shorter, bond-representing dowels. Mark the four long dowels 1.2 cm from each end, and then sharpen each end. Leave the other dowels flat at their ends. On each long dowel, place additional marks 10.6 cm from each previous mark, as shown at left of Fig. 89.

To assemble the layers, push the four long dowels completely through the four marked spheres of layer 2, until these dowels are centered, as shown at left of Fig. 89. Then push layers' 1 and 3 marked spheres into the sharpened ends of these long dowels, up to the marks. Dabs of glue where dowel enters sphere will help secure things. Be sure, before glueing, that the resulting pattern is identical with the one shown in Fig. 88. Then, and only then, place the eight shorter dowels between the marked spheres, in the planes of layers 1 and 3, as shown in Fig. 88. A dab of glue where flat end of dowel contacts sphere will hold it permanently in place. Position as accurately as possible (see greater detail of Fig. 89 right) before glueing, but do not push these shorter dowels into the spheres. Your finished model should resemble the one at left in the Fig. 90, which is the model of Fig. 88, turned upside-down.

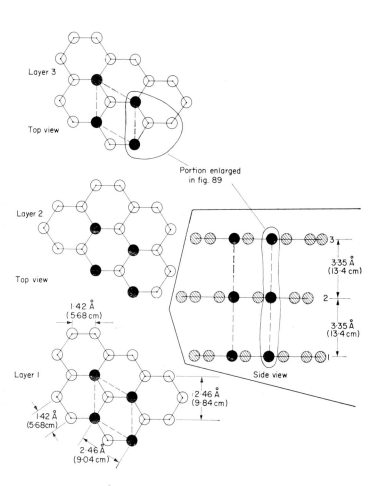

FIG. 88. Top views (left) and side view (right) of lattice model of graphite. Dark spheres and dotted lines show unit cell. Suggested scale represents a linear magnification of 4×10^8. The solid line encircled portions are enlarged in Fig. 89.

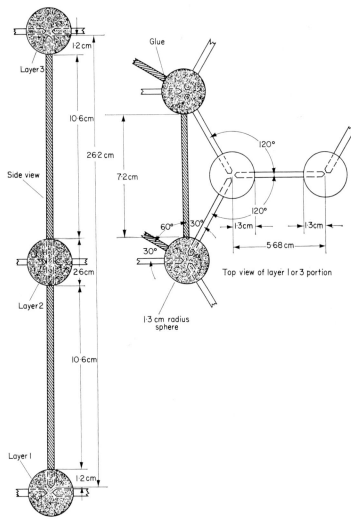

Fig. 89. Dowel detail of graphite lattice model. Dark spheres and dowels represent unit cell. Side view (left) and top view (right) represent enlargements of encircled portions of complete structure shown in Fig. 88.

LATTICE MODEL OF THE DIAMOND ARRANGEMENT

The rarer, but far more glamorous form of carbon, is of course, diamond. Figure 90 contrasts the diamond (right) and graphite structures, vividly showing the three-dimensional bonding array that makes diamond so hard. Note, in the diamond model of Fig. 90, that each sphere (representing a carbon atom) is surrounded tetrahedrally by four neighbors (see Fig. 91). Note also that some of the spheres at the corners of the unit cell in diamond (outlined by dark dowels) are not bonded to any spheres in that unit cell, but only to spheres outside it, that would, of course, be in neighboring unit cells. Note, for example, the sphere at the lower right hand corner of the unit cell shown at right of Fig. 90. It is bonded to spheres below, and in front of and to the right of, the planes defining the unit cell cube. If we intend to build only the unit cell portion of such a model, as shown in Fig. 92, bottom, this means we must have unit cell dowel rods to hold such spheres in place.

Although the unit cells shown in Figs. 92 and 94 are "upside-down" with reference to the one shown at right of Fig. 90, this is why four of the spheres in the Fig. 92 unit (numbered 15, 16, 17 and 18) have no bonds at all, and need the long, unit cell dowels to hold them. While it is perfectly true, in an extended model, that these spheres are each bonded to four other spheres, their bonds all lie outside the unit cell, and so are not shown in the limited model. To prepare this limited lattice model, on the same scale as the graphite model, select eighteen 1.3 cm radius spheres, and number them with a soft pencil as shown in Fig. 92. These numbers, which may be erased later on, will help you locate each sphere during the assembling of the model.

Spheres 1, 2, 3 and 4 are punched in a jig of pattern shown in Fig. 65(a) but with a 1.3-cm radius cut-out circle, as follows: center the sphere in the jig, punch along the "*A*" and "*C*" directions (see Fig. 65a), then withdraw these punches. Keeping the sphere centered in the jig, rotate it 90° about its vertical ("1–2" in the figure) axis, and then punch along the "*E*" and "*G*"

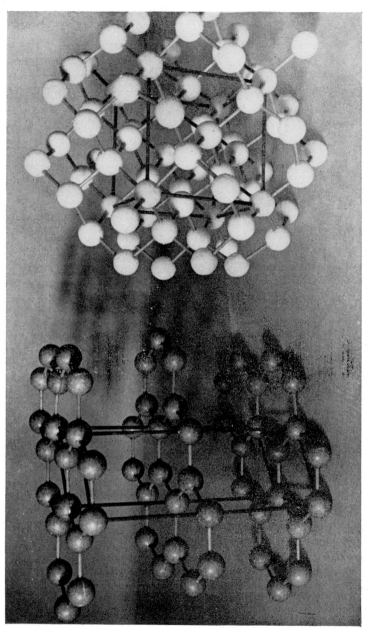

FIG. 90. Lattice models of diamond (right) and graphite (left), emphasizing two-dimensional layer structure of the latter,

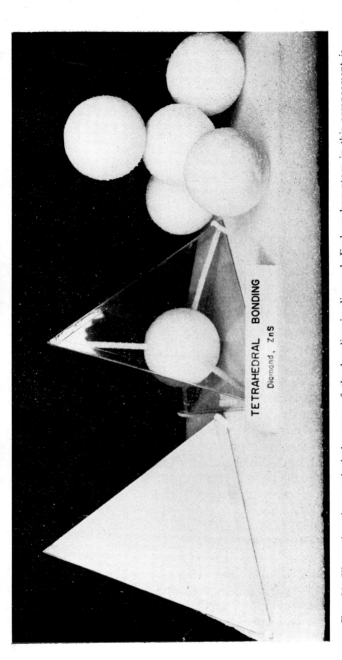

Fig. 91. Illustrating the tetrahedral nature of the bonding in diamond. Each carbon atom in this arrangement is surrounded by neighbors (right) so that the bonds extend (center) to the vertices of a regular tetrahedron (left).

Atomic and Molecular Models

directions. Spheres 5, 6, 7, 8, 9 and 10 are punched on the same jig with "*A*" and "*C*" punches only. The remaining spheres, numbers 11 through 18, are punched on an octahedral, NaCl-type jig (Fig. 65b) with a 1.3-cm radius cut-out circle: first, along all four directions shown, then, withdrawing any two opposite punches, and rotating the sphere 90° about the still-inserted ones, reinserting the two previously withdrawn punches. Of the six holes thus made, mark only three on one side, and ignore the other three. Finally, for each of spheres 11, 12, 13 and 14, swing an arc 1.2 cm radius from each marked hole (Fig. 93, top right), and at the intersection of these arcs, punch a hole half-way through the sphere to its center. Use the "wobble" test (Fig. 11) as you do this.

For bonds, cut sixteen lengths of dowel, each 6.16 cm long, and mark each 1.3 cm from each end. Cut twelve 14.24-cm lengths marking each of these 1.3 cm from each end, and then sharpen all ends of all dowels. Paint the longer dowels so as to distinguish them from the shorter, bond-representing ones, and assemble the spheres as shown in Fig. 92. Insert all dowels only part way up to their marks, and use no glue until later. When model begins to resemble the one in the middle of Fig. 94, try pushing all dowels in up to their marks, and see if the result is a fairly symmetrical cube. If not, push some of the long dowels past their marks, and pull others lightly out, until you get the squared off appearance of the cube. Then, making new marks on your long dowels, if necessary, to indicate their final positions, you may dab glue where each dowel enters each sphere to give a more permanent model. The dowel detail is shown in Fig. 93.

If desired, an approximate packing model of diamond, like the one shown at left of Fig. 94, may be constructed. It is approximate because a true packing model would show all spheres penetrating all other neighboring spheres. The amount of slicing required for this is frankly not worth the trouble. If you attempt this model, the use of 1.9-cm radius spheres is recommended. They can be numbered and punched exactly as for the lattice

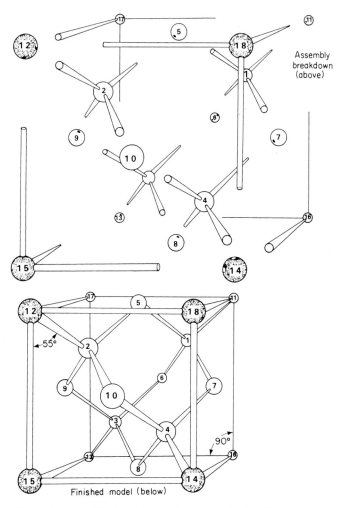

FIG. 92. Assembly breakdown (top) and final finished model (bottom) of lattice model of diamond unit cell. Spheres 1–4 take all four tetrahedral jig punches, spheres 5–10 take "*A*" and "*C*" punches only (Fig. 65a). Spheres 11–18 take three octahedral punches (each at right angles to the others), while spheres 11–14 also take one additional punch in middle of other three.

Dowels are inserted in spheres 1–4, and are linked together by spheres 5–10. Spheres 11–14 are then added. Spheres 15–18 are fitted with the long, white, unit cell—representing dowels, which are then fitted into the holes of spheres 11–14 as shown here.

161

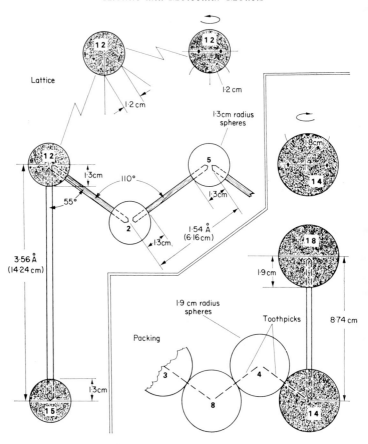

FIG. 93. Dowel detail for lattice model (upper left) and approximate packing model (lower right) of diamond unit cell. Long white dowels, and shaded spheres, represent unit cell outline.

model, except for the larger cut-out circles required for their jigs. Use toothpicks in place of short dowels, and twelve 8.74-cm-long dowels, marked 1.9 cm from each end, and sharpened at each end, for the unit cell dowels. In preparing spheres 11–14 for their final hole, swing arcs of 1.8 cm from each marked one (top right of

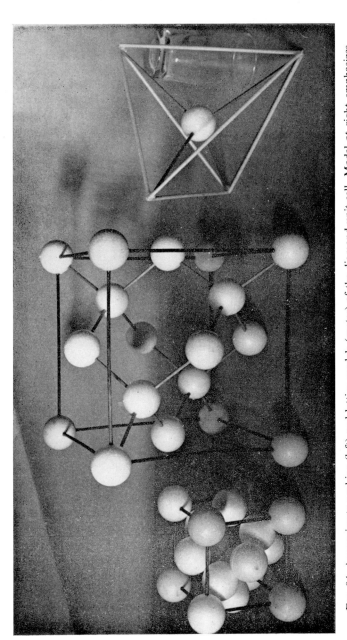

FIG. 94. Approximate packing (left) and lattice models (center) of the diamond unit cell. Model at right emphasizes tetrahedral nature of bonds between carbon-representing spheres.

Fig. 93). Assembly is identical to that for the lattice model and the finished product should resemble model at left of Fig. 94. Its dowel and toothpick detail is given at right of Fig. 93.

To emphasize the tetrahedral bond directions, you might prepare an extra sphere as you did numbers 1–4, insert four 6.16-cm-length dowels in it, and join the free ends of the dowels with ordinary drinking straws (you will need four 10-cm lengths), glueing them to dowels, and giving model shown at right of Fig. 94. The straws nicely outline the imaginary tetrahedron made by the centers of four surrounding, neighbor spheres.

LATTICE MODEL OF THE ICE-1 UNIT CELL

The final arrangement whose model construction will be formally described is that of ordinary ice, H_2O. It is a natural to follow the diamond arrangement because the oxygen-representing white spheres in its model (Fig. 97) show the same arrangement as the carbon-representing spheres in the diamond model (Fig. 94). A close look at the model photo in Fig. 97 will show that, although its unit cell is at a 90° orientation to the one shown in the center of Fig. 94, the way in which the white spheres are joined is identical. The only difference is that a small, dark sphere lies between each pair of white spheres in the Fig. 97 ice model. These small spheres, of course, represent hydrogen atoms, and each unit of white and two touching dark spheres represents an H_2O molecule. The model clearly shows the hydrogen bonding between the hydrogens of one water molecule and the oxygen of neighbor molecules.

As in diamond, and for the same reason, four of the corner "molecules" are not bonded to any neighbors within the given unit cell, and so must be held by the unit cell outlining dowels instead. It is suggested that eighteen 3.8-cm diameter white spheres represent the oxygen, and thirty-six 1.9-cm (about $\frac{3}{4}$-in.) diameter spheres, painted blue, represent the hydrogen atoms. The small spheres may be dipped in blue water-soluble paint, and allowed to dry, before fabrication. To fabricate them, simply punch a single hole through the center of each sphere, and clean through the other side. Use the "wobble" test (Fig. 11) when doing this.

The white spheres are numbered (see Fig. 95), and spheres 1–10 are punched (on the proper sized Fig. 65a-type jig) exactly as were spheres 1–4 in the diamond model. Spheres 11 through 18 are punched on the proper-sized octahedral (Fig. 65b) jig, exactly as their similarly numbered counterparts were for the diamond model. Cut sixteen 11-cm lengths of dowel for the bonds, and mark at both 1.9 and 2.85 cm from each end. Finally sharpen each end. For the unit cell outlines, cut twelve 25.44-cm lengths of dowel, and mark 1.9 cm from each end before sharpening the ends. Paint these long dowels so as to distinguish them from the shorter, bond-representing dowels.

To assemble the model, slide one small, blue sphere about to the middle of each short dowel rod, and then insert these dowel rods into white spheres 1–4, as shown in Fig. 95. The exact positions of the hydrogen-representing blue spheres is not known, and is thought to be constantly changing anyway, so the positions shown in Figs. 95 and 97 are not to be taken too literally. In fact, the symmetry has been deliberately destroyed in the Fig. 97 model, at the upper left front and rear corner spheres, to emphasize this point. You may use the Figs. 95 or 97 positions for the blue spheres, or some other better suited to your taste. The only requirement is that each white sphere be touched by only two small blue spheres, and be "hydrogen bonded" (within the model) to blue spheres on neighboring white ones.

In any case, when you decide on the placement of the blue spheres, slide them over to the proper white sphere, glueing them there. The purpose of the double marks on each short dowel (see the detail in Fig. 96) is to show how far to push both white and blue spheres on a dowel. Spheres 11–14 are given their final punch just as these numbers were in diamond. Swing 1.8 cm each octahedrally punched marked hole (Fig. 96 bottom right) and punch the final hole (towards sphere's center) at the intersection of these arcs. The remaining blue spheres can be attached by toothpicks, and positioned by eye to be at the approximate

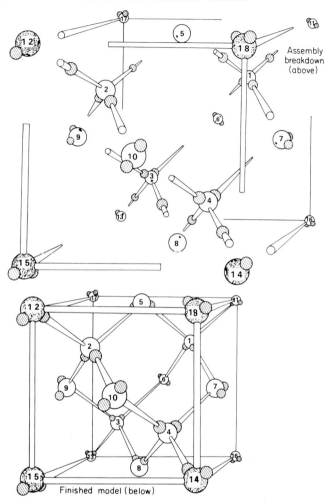

FIG. 95. Assembly breakdown (top) and final finished model (bottom) of lattice model of ice unit cell. Follow Fig. 92 instructions for spheres 1–18, but place small dark spheres (representing H atoms) on dowels and spheres as shown. Remember each white, oxygen-representing sphere has two "H atoms" touching it, while interior spheres also have two "H-bonds" where dowel goes direct from it through the "H atoms" of a neighboring "H₂O" unit.

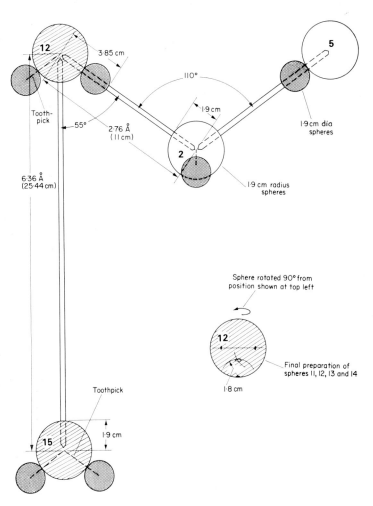

FIG. 96. Dowel detail of lattice model of ice. Sphere numbers correspond to those used in Fig. 95.

Fig. 97. Model of unit cell of ice. White spheres represent oxygen, dark spheres, hydrogen atoms. A single water molecule is represented by a white sphere with two dark spheres touching it. Long dowels represent imaginary unit cell outline.

tetrahedral angle to each other. The final assembly of long and short dowels is as in diamond. No glueing until the model appears well squared off, and resembles the one in the Fig. 97 photo.

SUMMARY OF CRYSTAL MODELS DESCRIBED IN UNIT 7

We shall now summarize the several structures whose model construction has just been detailed, including such things as number, size and suggested color, of the required spheres, and a partial listing of the substances each model represents. For a more complete listing, see the Appendix.

Close-packed metals. To build four layers, that can be positioned to represent either cubic or hexagonal close-packing, the packing model requires sixty-four $1\frac{1}{2}$-in. (4 cm) diameter unpainted spheres. An additional fourteen such spheres would be required to make the cubic unit cell. Such metals as Ag, Al, Au, Ca, Cu and Pb show cubic close packing, while Be, Cd, Cr, Mg and Zn crystallize with the hexagonal close packing arrangement. Some metals, such as Co and Ni, are known to form both arrangements. No lattice model for these structures was described.

Sodium chloride (rock salt) arrangement. Packing model requires thirteen $1\frac{1}{4}$-in. (3.2 cm) diameter dark-painted spheres to represent Na^+, and fourteen $2\frac{3}{8}$-in. (6 cm) diameter unpainted white spheres for the Cl^- ions. The lattice model needs are for thirteen $1\frac{1}{4}$-in. (3.2 cm) diameter dark-painted spheres for Na^+, and fourteen $1\frac{1}{2}$-in. (3.8 cm) diameter unpainted white spheres for Cl^- ions. This structure is shown by all alkali metal halides except CsCl, CsBr and CsI, as well as all silver halides except the iodide, and most alkaline earth oxides and sulfides.

Cesium chloride (body-centered cubic) arrangement. Packing model needs eight $2\frac{3}{8}$-in. (6 cm) diameter dark-painted spheres for Cs^+, and eight $2\frac{3}{8}$-in. (6 cm) diameter unpainted white spheres for Cl^- ions. Lattice model uses eight $1\frac{1}{2}$-in. (3.8 cm) diameter dark-painted spheres for Cs^+, and eight $1\frac{1}{2}$-in. (3.8 cm) diameter unpainted white spheres for Cl^- ions. The approximate, extended packing model described (Fig. 82) requires eighteen 2-in. (5.1 cm) diameter dark-painted spheres for Cs^+, and forty-eight $2\frac{3}{8}$-in. (6 cm) diameter unpainted white spheres for Cl^- ions. The compounds CsCl, CsBr, CsI, NH_4Cl and NH_4Br show this arrangement (the NH_4^+ ions represented by dark spheres), as well as numerous intermetallic compounds. If all spheres were the same size and color, the model would represent the body-centered cubic arrangement shown by the alkali metals, α-iron, Ba, V, W and, sometimes, Cr.

The calcium fluoride (fluorite) arrangement. Packing model needs fourteen $1\frac{1}{2}$-in. (3.8 cm) diameter dark-painted spheres for Ca^{++}, and eight 2-in. (5.1 cm) diameter unpainted white spheres for F^- ions. Lattice model requires fourteen $1\frac{1}{4}$-in. (3.2 cm) diameter dark-painted spheres for Ca^{++}, and eight $1\frac{1}{2}$-in. (3.8 cm) diameter unpainted white spheres for F^- ions. The alternate packing model described (Fig. 86) requires four $1\frac{3}{8}$-in. (3.5 cm) diameter dark-painted spheres for Ca^{++}, and twenty-seven 2-in. (5.1 cm) diameter unpainted white spheres for F^- ions. This so-called fluorite arrangement is shown by the fluorides of Ba, Ca, Cd, Cu, Hg, Pb, Sr; such oxides as CeO_2, HfO_2, ThO_2, UO_2, PbO_2 and ZrO_2; also by (except for Cs) the alkali oxides and sulfides, where the alkali ions would be represented by the white unpainted spheres in what is known as the anti-fluorite structure.

Graphite lattice structure. To make the three layers included in the unit cell, requires fifty-seven 1-in. (2.6 cm) diameter un-painted white spheres to represent the carbon atoms. Although some silicates and mica show layer structures, their layer details are not identical to this model.

Diamond unit cell. Lattice model needs eighteen 1-in. (2.6 cm) diameter unpainted white spheres for carbon atoms. The suggested approximate packing model requires eighteen $1\frac{1}{2}$-in. (3.8 cm) diameter unpainted white spheres for carbon atoms. This represents the structure of the diamond form of carbon, as well as silicon, germanium and gray tin. If every other sphere was dark painted, so these would alternate with the white ones, the model would represent the zinc blend arrangement shown by such compounds as CuF, CuCl, CuBr, CuI, AgI, AlP, SiC, and the sulfides of Be, Zn, Cd, Hg.

Ice unit cell lattice model. This model requires thirty-six $\frac{3}{4}$-in. (1.9 cm) diameter blue-painted spheres for hydrogens, and eighteen $1\frac{1}{2}$-in. (3.8 cm) diameter unpainted white spheres for oxygen atoms.

Summary of Sphere Requirements for all Above Models:

Sphere diameter		No. of spheres	Suggested colors of spheres:
(in.)	(cm)		
$\frac{3}{4}$	1.9	36	blue
1	2.6	75	unpainted white
$1\frac{1}{4}$	3.2	40	dark
$1\frac{3}{8}$	3.5	4	dark
$1\frac{1}{2}$	3.8	88	22 dark, others unpainted white
2	5.1	53	18 dark, others unpainted white
$2\frac{3}{8}$	6.0	78	8 dark, others unpainted white

SOME SUGGESTED USES OF CRYSTAL MODELS

Introduction

Those readers who have conscientiously expended the time and effort required to construct some of the models previously described, may well ask whether it was worth all the fuss. In view of our continuing concern about the nature of a model—as something imagined, rather than a copy of reality—this is a valid question indeed! If atoms and ions exist only in the imagination (for no one has actually seen them), what gives us the right to represent them with spheres of definite sizes and positions? The only answer—and it is a general one that might apply to any scientific model—would be that the model is useful in the sense that it explains, and often predicts, directly observable properties of the thing being modeled. Let us see if, and how, this applies to the crystal models whose construction was so carefully detailed in Unit 7.

The flat faces of crystals, the definite angles at which these faces meet (characteristic for each kind of crystal), and properties such as cleavage, are all nicely explained by the model of spheres put together in a systematic way. The hardness of diamond versus the slipperiness of graphite, although they are both arrays of carbon atoms, are likewise made understandable with the aid of their very different models (Fig. 90). These are qualitative things, well known enough to need no further mention here, and they are explainable in many possible ways by many possible models. The only requirement is that these models show the order that these properties suggest. Both the NaCl (Fig. 73) and CsCl (Fig. 82) models are orderly arrangements of spheres, so why model them with such different structures? A model worth its salt (no pun intended) must be quantitatively, as well as qualitatively useful. Let us therefore turn our attention to the quantitative differences in models such as NaCl and CsCl.

THE PROBLEMS OF DETERMINING ATOMIC SIZES

Let us look carefully at the NaCl and CsCl models just mentioned, for some clue about the origin of their different arrangements. Since packing models give a better idea of the imagined relative sizes of the ions involved, look at those shown at the left in Figs. 72 and 78. Since the differently colored spheres represent oppositely charged ions, it makes sense (by the laws of electrostatics) for each positive ion (dark sphere) to be surrounded by as many negative ions (white spheres) as possible—so long as these negative ions do not crowd each other too much. Note that because the "Na⁺ ion" in Fig. 72 is small, there is room without crowding for only six of the larger "Cl⁻ ions" around it. The "Cs⁺ ion," on the other hand (Fig. 78), is large enough to accommodate eight "Cl⁻ ions" with ease. Thus it is the imagined sizes of the ions involved, as represented by the different diameter spheres, which seem to be responsible for the two distinctly different models of these two otherwise closely related compounds.

Facing the question of how one measures the sizes of these imagined ions we cannot see is, first of all, one of proving that it is possible to measure the thickness of something without ever placing a ruler across it. Surely it is only indirectly that one can ever hope to measure anything as invisible as atoms and ions. The proof is readily available in the form of a large quantity of identical glass beads, which are found to occupy a definite volume when poured into graduated cylinders (Fig. 98). Is it possible to measure the thickness of one of the beads, indirectly, without ever bringing a ruler near the bead? The answer to this (Fig. 98) is to note that when the initial 150 cm³ of beads is transferred to a

FIG. 98. Illustrating the indirect determination of the diameter of a small glass bead. A certain volume of these beads, as measured in graduated cylinders (left), is found to occupy a given area (center) when spread into a layer one bead thick. The diameter thus indirectly calculable may be checked by direct measurement with a

flat tray, with an area of 500 cm², they make a single, complete layer, one bead thick. We may thus equate the initial 150 cm³ volume with the final volume, which can be represented by the expression 500 cm² × d cm, where d represents the required thickness of a single bead. The equation is justified because both volumes are for the same number of beads, packed together in the same, close, way. Thus

$$150 \text{ cm}^3 = 500 \text{ cm}^2 \times d \text{ cm}, \quad \text{and} \quad d = \frac{150 \text{ cm}^3}{500 \text{ cm}^2} = 0.30 \text{ cm}$$

the bead thickness we are seeking. The proof of this is simply to now use the ruler (right of Fig. 98), and thus confirm the indirect measurement.

This bead problem is a very good model indeed of the famous oil-slick experiment, first done by Lord Rayleigh in 1899,* and repeated many times since. Here, a single oil drop, made up of many oil molecules, and represented by the beads in the graduated cylinders, is allowed to spread out into a circular slick on water. Lycopodeum powder previously sprinkled on the water will show the size of the slick. This slick, thought to be one molecule thick, is represented by the beads when spread out in a single layer on the tray. If, as was found in a recently televized version of this experiment,† an oil drop $\frac{2}{100}$ in. in diameter spreads into a circular slick 8 in. in diameter, one can simplify the calculation by representing the drop as a cube $\frac{2}{100}$ in. on a side, and the slick as a rectangular solid 8 in. by 8 in. by d in., where d represents the slick's thickness. Assuming as before a constant volume throughout (i.e. that the oil does not dissolve in the water and has the same close arrangement of molecules in both drop and slick), and the above approximation to be reasonable in view of the lack of precision of the experiment, we may equate the volume of the

* Rayleigh, *Phil. Mag.* **48**, 337 (1899).

† These values are due to Prof. Eric Rogers, of Princeton University. They come from an experiment included in lesson 50 of Continental Classroom's "Atomic Age Physics" course, first shown on television in the U.S. on December 16, 1958.

cube to that of the rectangular solid. Thus, $(\frac{2}{100}$ in.$)^3 = $ (8 in. $\times$ (8 in.) $\times$ (d in.), or

$$d = \frac{(8 \text{ in.}^3)}{(8 \times 8 \times 10^6 \text{ in.}^2)} = 12.5 \times 10^{-8} \text{ in.}$$

This is roughly 30×10^{-8} cm, or, expressing it in units of atomic dimensions, about 30 angstroms (Å). An oil molecule is believed long, as molecules go, so we would expect most molecules to be perhaps an order of magnitude smaller than this.

As we have already indicated, proof of the existence of atoms (and hence calculation of their sizes) is based on indirect evidence. The Avogadro number N, for example ($N = 6 \times 10^{23}$), must be accepted as the number of atoms (or molecules) present in 1 gram-atomic (or gram-molecular) weight of any substance, once the arbitrary choice of a relative atomic weight scale is made. Although development of this concept, as well as various experimental determinations of N, is adequately covered in many texts,* one approach—a brief description of the way a radioactive element of known half-life may be used to find N—is readily understandable and especially convincing to beginning chemists. Here a suitable detector "counts" every one of the disintegrations of a known weight of the element, where each "count" represents the decay of one atom. Since one-half of the sample's atoms will decay during its half-life, a simple proportion yields the actual number of atoms in any weighable amount of the material.

Once N is known, the size of a substance's atoms or ions is calculable from the measured density—which may actually be determined in the laboratory, or easily looked up;† and from a knowledge of how these atoms or ions are packed together in the solid state. For an element whose atoms are packed together in a simple, symmetrical way (a majority of them do indeed show such an arrangement), the gram-atomic volume (space occupied by

---

* An excellent presentation may be found in *General Chemistry*, 3rd edition, by John A. Timm, McGraw-Hill Book Co., Inc., New York, 1956, pp. 46–55.

† *Handbook of Chemistry and Physics*, Chemical Rubber Publishing Co., Cleveland, Ohio.

one gram-atomic weight) is first found by dividing the gram-atomic weight (g) by the density (g/cm³). This in turn is divided by *N*, the total number of atoms present, to reveal the volume of a single atom. If this volume is treated as that of a cube, taking the cube root gives the approximate diameter of the atom itself—since it may be pictured as a sphere inscribed in this cube. As an example, consider finely powdered copper metal just because one gram-atomic weight of it (63.6 g) may so readily be "poured" into a graduated cylinder. Thus its gram-atomic volume may be read directly from the graduate, or calculated as indicated above. Since the diameter of any "simple" packed atom

$$= \sqrt[3]{\left(\frac{\text{g-atomic wt.}}{\text{density} \times \text{N}} \right)},$$

diameter of the Cu atom

$$= \sqrt[3]{\left(\frac{63.6 \text{ g}}{8.96 \text{ g/cm}^3 \times 6 \times 10^{23}} \right)} = 2.28 \times 10^{-8} \text{ cm.}$$

Its radius is therefore 1.14×10^{-8} cm, or, more conveniently, 1.14 Å, since 1 Å $= 10^{-8}$ cm. The 10% error between this and the experimental value given in Table 3 is due, in part, to the assumption that an atom may be treated as a cube.

The experimental values for atomic radii were obtained by halving the internuclear distances between the atoms, as revealed by X-ray diffraction experiments. In such experiments, a beam of X-rays is directed at the atomic planes of a crystal at a known angle, θ. Figure 99 shows this in cross-section for a sodium chloride type crystal. Notice that the bottom X-ray wave must travel a total of $2d \sin \theta$ more than the top wave to reach a film placed at the top right of the figure. If this distance corresponds to one, two, or three, etc., wavelengths of the X-rays being used, then both waves will reinforce each other, and so expose the film. From examination of such exposed portions of films, it is possible to deduce the distances between the layers of atoms in a crystal. The interested reader is referred, for further details, to the sources listed in the appendix.

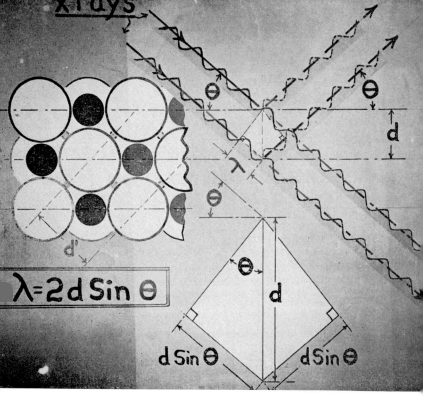

Fig. 99. Derivation of the Bragg equation (lower left). If portion of incident X-ray beam (top left) is assumed to be "reflected" (top right), the lower ray must travel a total of $2d \sin \theta$ farther than the upper one. For these two "reflected" rays to reinforce each other, this distance must equal $n\text{-}\lambda$, where $n = 1, 2, 3. \ldots$ Here, d (far right) represents the M–X distance in an NaCl-type crystal shown at far left.

178

An alternate approach to determining the radii of metal atoms such as copper involves our close-packed model because copper crystallizes with the so-called cubic close packed arrangement. As previously mentioned, the unit cell of this arrangement is a

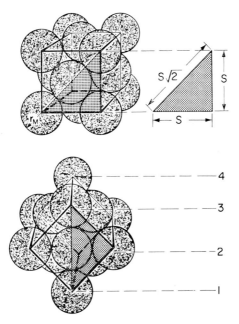

FIG. 100. *Top:* unit cell of the cubic close packed arrangement. Note that only the diagonal of a face (hypotenuse of the shaded triangle) is directly the sum of atomic radii (r_M), and is equal to $4 \, r_M$. *Bottom:* the unit cell as it sits in the four layers of the cubic close packed model described in Figs. 66–69. Tilted over at a 45° angle, it extends through all four layers, numbered 1, 2, 3, 4 in the drawing.

tilted over cube (Fig. 100) which extends through all four layers of our model. Since density is the weight of a unit volume of a substance, we may let this known density equal the weight, divided by the volume, of this tiny unit cell. The weight is the combined weight of all atoms belonging to the unit cell, but it must be remembered that this is not simply the weight of its fourteen atoms, since each of these is shared by neighboring unit cell cubes (Fig. 101). Thus the eight corner atoms ("*A*" in Fig. 101) are shared by eight cubes; the six in the centers of the faces

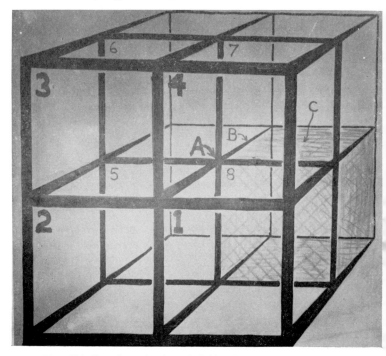

FIG. 101. Drawing of cube subdivided into eight smaller cubes, designed to show that sphere whose center is at "*A*" is shared by all eight cubes. Sphere with its center at point "*B*" is shared by four of the cubes (nos. 5, 6, 7, 8); while one with center in face of cube 8 (point "*C*") is shared by two cubes (nos. 7 and 8).

("*C*" in the figure), by two cubes; those on an edge ("*B*" position), although there are none such here, by four cubes. Our unit cube therefore "owns" $8 \times \frac{1}{8}$, or 1 corner atom, and $6 \times \frac{1}{2}$, or 3 others: a total of 4 atoms. Hence the weight of our cubic unit cell is 4 times the weight of 1 atom.

The weight of a single atom is calculable if we assume the chemists' atomic weight scale, and his model of the Avogadro number, N. Since one gram-atomic weight of any element, or 1-gram-molecular weight of any compound, is believed to contain this number of particles, a single particle atom must have $1/N$th of this weight. The accepted value of N as 6×10^{23} means that a copper atom, for example, would weigh

$$\frac{63.6 \text{ g/mole}}{6 \times 10^{23} \text{ atoms/mole}}, \text{ or } 1.06 \times 10^{-22} \text{ g/atom}$$

The volume of our unit cell cube may be expressed in terms of the desired atomic radius, r, because the diagonal of a cube face is 4 times r (Fig. 100, top), as well as side, $s \times \sqrt{2}$, by the Pythagorean theorem. Thus $4r = s\sqrt{2}$, and side $s = 2r\sqrt{2}$. Volume, V, in turn, $= s^3 = (2r\sqrt{2})^3 = 16\sqrt{2}\,r^3$. Now if the known density

$$D = \frac{\text{weight (g)}}{\text{volume (cm}^3)},$$

which in turn equals

$$\left(\frac{\dfrac{4 \times \text{gram-molecular weight (GMW)}}{N}}{16\sqrt{2}\,r^3} \right),$$

we can solve this expression for r, obtaining

$$r = \sqrt[3]{\left[\frac{\dfrac{4 \times \text{GMW}}{N}}{16 \times \sqrt{2} \times D} \right]} = \sqrt[3]{\left(\frac{4 \times \text{GMW}}{6 \times 10^{23} \times 16\sqrt{2} \times D} \right)}$$

$$\sqrt[3]{\left(\frac{.29 \times \text{GMW}}{D} \times 10^{-24} \right)}.$$

For copper, with a density of 8.96 g/cm³,

$$r = \sqrt[3]{\left(\frac{.29 \times 63.6}{8.96} \times 10^{-24} \right)} = 1.27 \times 10^{-8} \text{ cm},$$

in agreement with the accepted experimental value. Table 3 lists several atomic radii similarly determined, and shows their close agreement with experiment. Such agreement is convincing proof indeed of the validity of the cubic close-packed models, since all metals in that table are known to crystallize with that structure.

TABLE 3. COMPARISON OF CALCULATED AND EXPERIMENTAL ATOMIC RADII OF CUBIC CLOSE-PACKED METALS

Metal symbol	g-atomic weight (g)	Density[a] (g/cm^3)	Metallic radius (cm $\times$ 10^{-8})	
			calculated[b]	experimental[c]
Ag	108	10.5	1.44	1.44
Al	27	2.70	1.43	1.43
Au	197	19.3	1.44	1.44
Ca	40	1.55	1.96	1.96
Cu	63.6	8.96	1.27	1.27
Ni	59	8.90	1.24	1.24
Pb	207	11.3	1.75	1.74
Pt	195	21.5	1.38	1.38
Sr	88	2.6	2.14	2.13

[a] Lange's *Handbook of Chemistry*, 1956 values, rounded off to three places.

[b] Slide rule accuracy, using the formula

$$\text{radius} = \sqrt[3]{\left(\frac{.29 \times \text{g.-atomic wt.}}{\text{density}} \times 10^{-24} \text{ cm}^3 \right)}.$$

[c] From Wyckoff, *Crystal Structures*, Interscience, New York and London, 1948.

THE SODIUM CHLORIDE STRUCTURE

The type approach previously described is just as well applied to rock salt-type structures. Indeed, the face-centered cubic close-packed arrangement can be shown to be very closely related to such structures. To see just how, consider the close packed arrangement shown in Fig. 66, and note the three types of holes that exist between the spheres. In a given layer, there is what we might call a trigonal hole, bounded by the three spheres in that layer. Between layers, two other kinds of holes are present: one (Fig. 66), between a single first layer (light colored) sphere, and the three second layer (dark) spheres touching it; the other, adjacent to it, between three first layer, and three second layer spheres. If we consider the geometric center of each hole, and the centers of the spheres surrounding each, we might use the terms tetrahedral and octahedral, respectively, to describe these holes. The terms arise from the fact that connecting the centers of the spheres which surround the holes would produce the two solids these names come from: tetrahedron and octahedron.

The unit cell of the face-centered, cubic close packed arrangement, shown in Fig. 100, has thirteen of the octahedral type holes between its fourteen spheres: four each in the top and bottom layers, as the cube is oriented in the top drawing of the figure, and five in the middle layer, counting the one hidden in its center. To see why this cube is so closely related to NaCl, rock salt-type, structure, one need only place thirteen of the proper-sized dark spheres into those octahedral holes (Figs. 102 and 103), to see the cubic close-packed unit cell become the unit cell of the rock salt arrangement.

This rock salt arrangement was, in fact, worked out by Barlow* more than 15 years before anyone ever thought of passing X-rays through such crystals. His method is particularly appropriate to this discussion because it takes into account the external shape of the crystal, which is thought to be the result of the piling up of many sub-microscopic building blocks—each supposedly an exact copy in miniature of the parent crystal. Each building block, in turn (our so-called unit cell), is made of atoms or ions which are pictured as hard spheres, so that the problem boils down to how most efficiently to pack such spheres together. Thus the method is both simple and direct, and is, after all, just what one does when constructing packing models.

In the case of the cubic rock salt crystal, Barlow (correctly) deduced a face-centered cubic close packing of chloride ions, with

* W. Barlow, *Sci. Proc. Roy. Dublin Soc.* **8**, 527 (1893–8).

FIG. 102. Packing model (right) of portion of cubic closed packed metal structure. The unit cell (made from white spheres) is emphasized by omitting surrounding portions of top two layers. An enlarged model of this

Fig. 103. The large unit cell model of Fig. 102 (left) with its octahedral holes filled with dark spheres, showing its relation to the NaCl structure of Fig. 72. Small unit cell model, with octahedral holes empty, is shown at right for comparison.

sodium ions occupying the octahedral holes (Fig. 103). In other words, large spheres represent Cl^- ions, the smaller ones, Na^+ ions. Let us consider this, the structure shown by most alkali metal halides, and attempt to derive a formula for the internuclear distance (d_{M-X}) between oppositely charged ions. Since densities are readily available and easily understood, why not set the known density of an alkali halide equal to the weight divided by the volume of its tiny face-centered cubic unit cell?

The weight of such a cube is simply the combined weight of its ions, although one must remember when determining this weight that each corner ion is shared by seven other cubes, each edge ion by three others, and each face ion by one other (see Fig. 101). Thus one-eighth, one-fourth and one-half, respectively, of these ions belong to our unit cell cube, which, with the metal ion in its body center, make four of each kind, or four M–X "molecules". The cube's weight is therefore four times the halide's gram-molecular weight, divided by N, the number of "molecules" per gram-mole. Its volume is simply the side, S, cubed, where S is twice the M–X distance we wish to solve for. Thus

$$\text{density} = \frac{\text{weight}}{\text{volume}} = \frac{4 \cdot \dfrac{\text{GMW}}{6 \times 10^{23}}}{(2\, d_{M-X})^3}$$

or

$$d_{M-X} = \sqrt[3]{\left(\frac{0 \cdot 833\ \text{GMW}}{\text{density}}\right)} \cdot 10^{-8}\ \text{cm} \tag{1}$$

The results of such M–X distance calculations in rock salt type halides are found to be in good agreement with X-ray determinations (Table 4). Indeed, the former method was essential to the Bragg's X-ray work[*] because, when the Bragg equation $n\lambda = 2d \sin\theta$ was initially applied to the analysis of rock salt-type crystals, neither λ nor d were known. By finding d as we just

[*] W. H. Bragg and W. L. Bragg, *Proc. Roy. Soc.* A **88**, 428 (1913); and W. L. Bragg, *ibid.* **89**, 248 (1913).

have (it is the d_{M-X} of equation (1)), one could then calculate λ, and use X-rays of this wavelength to find d spacings in other NaCl-like crystals (see Fig. 99).

TABLE 4. INTERNUCLEAR (M–X) DISTANCES

A comparison of d_{M-X} values calculated using equation (1) with those obtained from X-ray crystallography experiments.

Compound	GMW (g)	Density [a] (g/cm^3)	d_{M-X} (Å)	
			calc'd	X-ray [b]
LiF	25.9	2.30	2.10	2.01
LiCl	42.4	2.07	2.57	2.57
LiBr	86.8	3.46	2.75	2.75
LiI	134	4.06	3.02	3.00
NaF	42.0	2.79	2.32	2.31
NaCl	58.5	2.16	2.82	2.81
NaBr	103	3.21	2.98	2.98
NaI	150	3.67	3.24	3.23
KF	58.1	2.48	2.68	2.67
KCl	74.6	1.99	3.14	3.14
KBr	119	2.75	3.30	3.29
KI	166	3.13	3.53	3.52
RbF	92.4	2.88	3.11	2.82
RbCl	121	2.76	3.31	3.27
RbBr	165	3.35	3.43	3.43
RbI	212	3.55	3.68	3.66

[a] Handbook values, rounded off to three places.
[b] From R. W. G. Wyckoff, *Crystal Structures*, Vol. 1, Interscience Publishers, New York, 1951.

DETERMINING IONIC RADII

Now that the determination of internuclear distances has been found amenable to simple development, the problem still remains of how much of this to assign the metal, and how much to the halide ion. A naïve approach would be to assume that these ions act like the hard spheres we have already usefully pictured, so that their radii are fixed and unaffected by the nature of their neighbors. The beauty of this admittedly questionable assumption is that we may attribute the change in M–X distances within a particular metal's halides completely to the change in size of the halide ions (see Fig. 104). In like manner, the d_{M-X} changes in alkali iodides, for example, are attributable only to the differing sizes of the metal ions (Fig. 105).

When one looks at the increases in M–X distances in such compounds (Table 5), one is struck with the essentially constant increase in size that comes from replacing one given ion with another. Except for lithium fluoride, for example, the M–X

TABLE 5. AN ANALYSIS OF INTERNUCLEAR DISTANCES

A study of the increments in M–X distances among rock salt-type alkali metal halides. Note the essentially constant changes, as represented by the average Δd_{M+} and Δd_{X-} values. The M–X distances themselves are the X-ray values from Table 3.

M^+ \ X^-	F^-	Δd_{X-}	Cl^-	Δd_{X-}	Br^-	Δd_{X-}	I^-	Avg. Δd_{M+}
Li^+	2.01	0.56[a]	2.57	0.18	2.75	0.25	3.00	↓
Δd_{M+}	0.30[a]		.24		0.23		0.23	0.23
Na^+	2.31	0.50	2.81	0.16	2.98	0.26	3.23	
Δd_{M+}	0.36		.33		0.31		0.29	0.32
K^+	2.67	0.47	3.14	0.15	3.29	0.24	3.52	
Δd_{M+}	0.15		.13		0.14		0.14	0.14
Rb^+	2.82	0.45	3.27	0.16	3.43	0.23	3.66	
avg $\Delta d_{X-} \rightarrow$	0.48		0.16		0.25			

[a] Not used to find average Δd values.

distances in sodium halides are all about 0.23 Å greater than the corresponding lithium salts. This may be interpreted to mean that the Na^+ ion's radius is 0.23 Å larger than that of the Li^+ ion. The potassium ion, using this type of argument, is 0.32 Å larger than is the sodium ion, and so on. In like manner it follows (from the M–X distances given in Table 5) that the average increase in size of halide ions, proceeding from F^- to I^-, is 0.48, 0.16 and 0.25 Å, respectively. These distances, given in

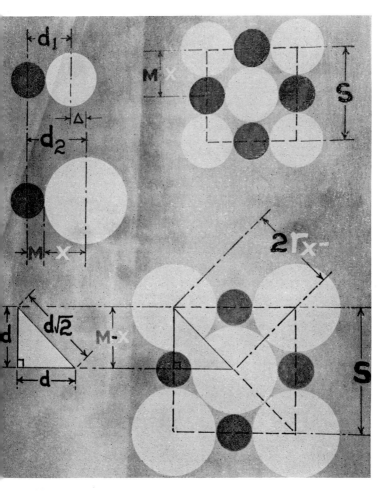

FIG. 104. Illustrating the increase in M–X distance due only to anion (top left), and relation of anion radius to M–X distance (lower left and right) that follows from assuming anions to be in contact (lower right).

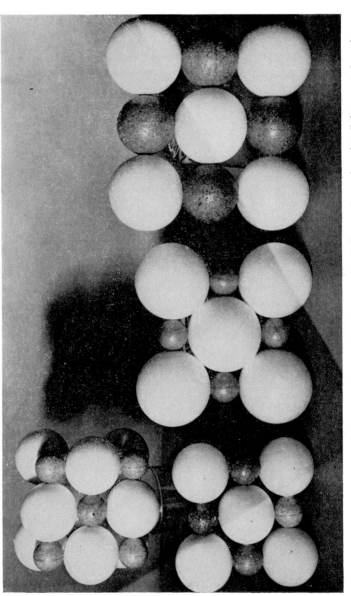

Fig. 105. Models showing increase in size of rock salt type unit cell due to change in anion (left to center), and cation (center to right). Model at left is the packing model of NaCl shown in Fig. 72, with outer layer removed

Table 5, and the use to which they are here being put, are due to Goldschmidt.* The value of this approach is simply that the fixing of any one metal or halide ion radius is all that is needed to fix all the other ionic radii. Our only task, therefore, is to find a reasonable way to determine one such radius, and it is to this problem that we now turn.

Perhaps the most straightforward approach comes from our picture of the rock salt structure itself. If it is indeed a cubic close packing of halide ions, with metal ions occupying the octahedral sites, perhaps a combination of large halide and small metal ions, in which the latter were just small enough to occupy these sites without pushing the halide ions apart, would justify considering the face diagonal of such a unit cell to be simply four times the halide ion radius (lower right of Fig. 104). It then follows, by the Pythagorean theorem, that $4r_{X^-} = 2\sqrt{2}\, d_{M-X}$, or that

$$r_{X^-} = 0.707\, d_{M-X} \qquad (2)$$

If it were not for the departure of Li from the pattern of d_{M-X} increments shown in Table 5, we might choose LiI as most likely of consideration. This is, in fact, the compound Landé† chose in his attempts to work out a set of ionic radii. Indeed, it is his approach we are using, although not all of our M–X distances were available to him. But because of lithium's odd behavior, we are led to suspect that its powerful field might pull the I^- ions so close together that they would overlap each other. The resulting internuclear distance might not then indicate the hard sphere I^- radius we are seeking.

The next most logical choice would be NaI, so let us assume that this is the case in which the I^- ions just touch each other in their face centered lattice. It now follows that the I^- radius (r_{I^-}) is related to the NaI internuclear distance as indicated by equation (2). Thus, $r_{I^-} = 0.707\, d_{NaI} = 0.707 \times 3.23\ \text{Å} = 2.28\ \text{Å}$; and

* V. M. Goldschmidt, *Skrifter Norske Videnskaps-Acad. Oslo*, **2**, 16 (1926).
† Landé, A., *Z. Physik*, **1**, 191 (1920).

since $d_{\text{NaI}} = r_{\text{Na}^+} + r_{\text{I}^-}$, $r_{\text{Na}^+} = 3.23 - 2.28 = 0.95$ Å. Finally, by our assumption that these ions retain their respective radii in all their compounds, we can quickly determine all other ionic radii by adding or subtracting the average $d_{\text{M-X}}$ increments given in Table 5. The Br$^-$ radius, for example, would be 2.03 Å (2.28 − 0.25); the Cl$^-$ radius, 1.87 Å (2.03 − 0.16), and so on. Metal ion radii are similarly calculable starting with the Na$^+$ radius we have just determined.

VALIDITY OF THE ABOVE APPROACH

That this naïve approach is also a reasonable one becomes apparent when one compares its results with those of more sophisticated calculations. Pauling,* for example, sought in 1927 to determine ionic radii from a quantum mechanical approach. Using the compound NaF as an illustration, one could qualitatively expect the Na$^+$ to be smaller than the F$^-$ ion. This follows because, while both ions exhibit the neon structure, they possess, respectively, one more, and one less nuclear charge with which to pull in their extra-nuclear electrons. But instead of assuming their radii to be merely inversely proportional to these nuclear charges, Pauling made use of the idea of a screening effect, due to the inner electrons, which stays constant for a given electron arrangement. For species with the neon arrangement, this screening constant is calculated from the quantum mechanics to be 4.5 It then follows that the effective nuclear charge felt by the "outer" electrons, and hence the one that determines ionic size, is the total charge, Z, minus this screening constant. Thus,

$\dfrac{r_{\text{Na}^+}}{r_{\text{F}^-}} = \dfrac{Z_{\text{F}^-} - S}{Z_{\text{Na}^+} - S}$, where s is the screening constant. This ratio of $\dfrac{4 \cdot 5}{6 \cdot 5}$, i.e. $\left(\dfrac{9 - 4 \cdot 5}{11 - 4 \cdot 5}\right)$, tells us that 4.5/11 of the 2.31 Å NaF inter-

nuclear distance (i.e. 0.95 Å of it) belongs to the Na$^+$ ion, while

* L. Pauling, *J. Am. Chem. Soc.* **49**, 765 (1927).

6.5/11 of it (1.36 Å) is due to the F⁻ ion. Similar calculations, using the appropriate inert gas screening constants, give the calculated values shown in Table 6. Note the approximate agreement between these and our naïve values. The advantage to beginners of the latter figures is that their derivations can be easily followed, while screening constants must be accepted more or less "on faith".

TABLE 6. IONIC RADII

A comparison of the radii determined by our naïve approach, with those calculated by Pauling.

M⁺ ion	Ionic radius (Å)		X⁻ ion	Ionic radius (Å)	
	naïve	calc'd		naïve	calc'd
Li	0.72	0.60	F	1.40	1.36
Na	0.95	0.95	Cl	1.87	1.81
K	1.27	1.33	Br	2.03	1.95
Rb	1.41	1.48	I	2.28	2.16

THE INFLUENCE OF IONIC RADII UPON CRYSTAL STRUCTURE

Back on p. 173 we saw that the differences in coordination—of sodium ions by six chloride ions, and of cesium ions by eight chlorides—were responsible for the different arrangements of these ions in sodium, and cesium chloride. A possible explanation for this difference was put forward: that it was the small size of the sodium, and larger size of the cesium ion, which accounted for their different coordination. Specifically, the laws of electrostatics suggest that, while each positive ion tries to surround itself (be coordinated) by as many negative ions as possible, the number of such negative ions might be limited by how many of them could contact the positive ion, while not crowding each other past the point of mere anion-to-anion contact.

Now admittedly, the assumption that ions can be represented by hard spheres, of unchanging radii, is questionable. But, having just seen how useful such an assumption was in determining ionic radii, it seems perfectly reasonable to follow this up by attempting to see whether or not the arrangements predicted by ionic sizes alone are in fact found in nature. Let us first turn to the problem of calculating the size of a sphere that could fit

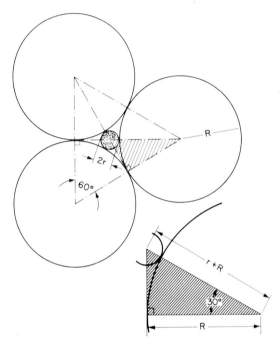

Fig. 106. Suggested approach to calculation of radius ratio of dark/white spheres. The white spheres may be thought of as a portion of a layer of close-packed metal atoms, so that the dark sphere represents the largest foreign particle that could enter this array without pushing its atoms apart.

snugly between three, four, six or eight larger spheres, contacting, but not pushing apart these larger, close-packed spheres. Since the three, four and six-sphere-surrounded holes exist in close-packed structures (see p. 182), we might rephrase the problem as one of finding the sizes of atoms that could penetrate, or reside in, close-packed metal crystals without disrupting their structures. Questions such as this are being considered in various research laboratories today.

Figures 106–110 show possible approaches to such calculations,

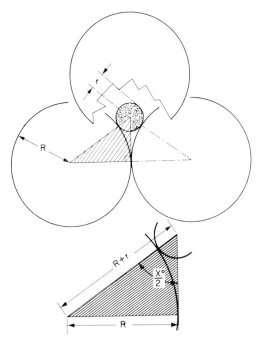

FIG. 107. Suggested approach to finding size of hole (represented by dark sphere) formed by four metal atoms (represented by white spheres) of any close-packed metal structure. Solution of shaded triangle (bottom) yields ratio of r/R. Key to solution is that angle $x/2$ is just half the tetrahedral angle of 110°.

and suggest that simple geometry considerations will yield their solutions. For the three-sphere-surrounded trigonal hole, the shaded triangle shown at top left of Fig. 106, and enlarged at bottom right of the figure, reveals that the hypotenuse of the shaded triangle is the sum of the small (r) and large sphere (R)

radii, while the longer leg is simply R. From the Pythagorean theorem we see that these are in the ratio $2/\sqrt{3}$, which gives us the proportion $(r + R)/R = 2/\sqrt{3}$. Solving this for the ratio of the small over the large sphere radius, or r/R, yields 0.155, which tells us the small sphere can have a radius up to 0.155 of the large spheres' radii without disrupting them.

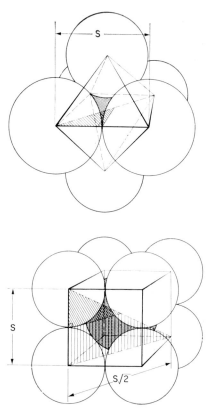

FIG. 108. *Top:* octahedral arrangement of white spheres around dark sphere (well hidden). Shaded triangle, lying in plane joining centers of four middle spheres, is the triangle shown in Fig. 109. *Bottom:* body-centered cubic arrangement of white spheres around dark sphere (well hidden). Small shaded triangle, lying in cube's diagonal plane, is one shown in Fig. 110, top left.

Equally straightforward calculations are possible for radius ratios of spheres that might occupy tetrahedral, octahedral, and body-centered cubic type holes. The tetrahedral case is shown in Fig. 107, where shaded triangle's solution is required. Here, since $X/2$ is half the tetrahedral angle (refer to Fig. 26b), or 55°, the sine of this angle, 0.82, can be set equal to the ratio of long leg/hypotenuse, which is $R/(r + R)$. From this relationship we find that $r/R = 0.22$, which says that the small sphere can have a radius up to 0.22 times its tetrahedrally surrounding large spheres' radii without forcing the latter apart.

Figures 108, top, and 109 suggest a similar calculation for an octahedrally surrounded sphere, using shaded triangle. Here we

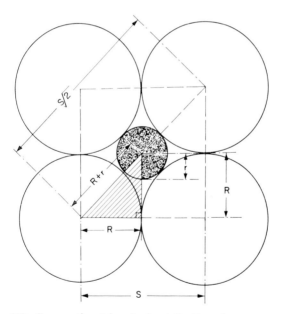

FIG. 109. Cross-section taken horizontally through arrangement shown at top of Fig. 108. Solution of small shaded triangle (bottom left here) shows maximum size of sphere that could occupy octahedral hole made by six white spheres of Fig. 108 top, without forcing those spheres apart.

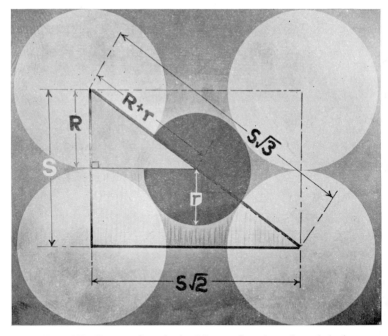

Fig. 110. Cross-section obtained by passing plane through body diagonal of cubic arrangement shown at bottom of Fig. 108. Solution of small shaded triangle shown here yields maximum size dark sphere could have to occupy hole formed by the eight white spheres shown at bottom of Fig. 108, without pushing these white spheres apart.

find that the ratio $R/(r + R) = 1/\sqrt{2}$, and that $r/R = 0.414$. Finally, for the body-centered cubic type "hole", surrounded by eight spheres in contact, note (Fig. 108, bottom) that it is the small, shaded triangle, lying in the outlined cube's diagonal plane, that yields the desired solution shown in Fig. 110. Note, by Pythagoras, that the sides of this small triangle (Fig. 110, top left) are in the ratio 2 to 1 to $\sqrt{3}$ for hypotenuse, short, and long side respectively. Since short side is R, and hypotenuse, $R + r$, we may write the proportion $R/(R + r) = 1/\sqrt{3}$. Solving for r/R gives 0.732.

FIG. 111. Packing models to illustrate how size of central (dark) sphere determines how many larger, white spheres may comfortably surround it. These models illustrate the coordination of a positive ion (dark sphere) by negative ions (white spheres) in: (left to right) the cesium chloride, sodium chloride, and zinc sulfide type structures.

The latter two ratios are of most direct concern, since they tell us that any positive ion whose radius is between 0.414 and 0.732 of the negative ion radius has room for six of the latter and is therefore expected to crystallize with the sodium chloride type structure. When the positive ion radius is more than 0.732 of the negative ion's, however, the former can accommodate eight of the latter, and the cesium chloride structure is to be expected

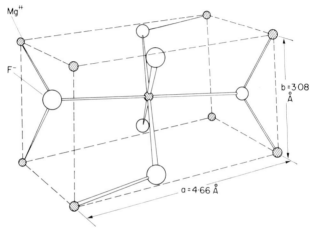

FIG. 112. Unit cell of the MgF_2 arrangement. This compound, an example of the so-called Rutile structure, is one in which each Mg^{++} ion (dark spheres) is surrounded octahedrally (or would be in an extended model) by six F^- ions (white spheres). Each F^- ion, on the other hand, is surrounded by three Mg^{++} ions in a triangular planar (trigonal) arrangement.

(see Fig. 111). If we examine the radius ratios of Na^+/Cl^- and Cs^+/Cl^-, we find them to be, respectively, $0.95/1.81 = 0.525$, and $1.69/1.81 = 0.934$, which is in agreement with our predictions.

When one examines the radius ratios of various ionic compounds, and attempts to predict from them whether the positive ion should be surrounded by six, or by eight, negative ions, such predictions are verified in a majority of cases. This is true not only for the sodium– versus cesium–chloride type structures, but also in the case of MX_2 type compounds such as CaF_2, fluorite, and MgF_2. As the former's model shows (Fig. 86), each calcium ion is surrounded by eight fluoride ions. In view of the radius ratio, $r_{Ca^{++}}/R_{F^-} = 1.00/1.36 = 0.736$, this was to be expected. MgF_2, on the other hand, whose structure is shown in Fig. 112 has each Mg^{++} surrounded by only six F^-. Here, the ratio of $r_{Mg^{++}}/R_{F^-} = 0.65/1.36 = 0.478$, which would predict this type of coordination.

Fig. 113. Packing model of uranium metal, according to the data supplied by Jacob and Warren in 1937. This is an example of the many additional models, not described in this book, which can easily be constructed from styrofoam spheres.

In Conclusion

Thus we see that both the models we can build, and the naïve assumptions we must make in the process, are useful in the sense that they explain, and even predict, things found to hold true for the substances being modeled. While we do not pretend that the substances actually look like the models, we can say with confidence that—in the scientific sense—they are useful models indeed.

While we have considered only a limited number of models and applications, there is no reason why the interested reader cannot look up almost any structure he wishes in the literature, and gain understanding of it by constructing its model. The packing model of uranium metal (Fig. 113), which shows a different arrangement than any we have modeled, is shown here as an example. If the reader is convinced that the building of such models is worth the effort, then this little book will have achieved its purpose.

APPENDIX

COMPLEX ION ARRANGEMENTS (incomplete listing)

Planar triangular: CO_3^{--}, NO_3^{--}

Pyramidal: ClO_3^-, IO_3^-

Tetrahedral: BF_4^-, PO_4^{---}, SO_4^{--}, ClO_4^-, $CoCl_4^{--}$, $AlBr_4^-$, CrO_4^{--}, $Be(H_2O)_4^{++}$

Square planar: $Ni(CN)_4^{--}$, $PtCl_4^{--}$, $Cu(NH_3)_4^{++}$, $PdCl_4^{--}$

Octahedral: $Fe(CN)_6^{---}$, $Fe(CN)_6^{----}$, AlF_6^{---}, $Al(H_2O)_6^{+++}$, $Cr(H_2O)_6^{+++}$

ATOMIC AND IONIC RADII ACCORDING TO PAULING AND WYCKOFF

Name	Symbol	Van der Waals or metallic	Co-valent	Ionic[a]	Source[b]
Aluminum	Al	1.43		0.50 (+3)	W
Barium	Ba	2.17		1.25 (+2)	W
Beryllium	Be	1.13		0.31 (+2)	P
Bromine	Br	1.95	1.14	1.95 (−1)	P
Calcium	Ca	1.96		1.00 (+2)	W
Carbon	C	1.7	0.77		P
Cesium	Cs	2.62		1.69 (+1)	W, P
Chlorine	Cl	1.80	0.99	1.81 (−1)	P
Chromium	Cr	1.35		0.70 (+3)	W
Cobalt	Co	1.25		0.78 (+2)	W
Copper	Cu	1.27		0.86 (+1) / 0.70 (+2) / 1.36 (−1)	W
Fluorine	F	1.35	0.64		P
Hydrogen	H	1.2	0.3	1.3[c] (−1)	W
Iodine	I	2.15	1.33	2.16 (−1)	P
Iron	Fe	1.24		0.80 (+2)	W
Lead	Pb	1.74		1.18 (+2)	W
Lithium	Li	1.51		0.60 (+1)	W
Magnesium	Mg	1.60		0.65 (+2)	P
Nickel	Ni	1.24		0.74 (+2)	W
Nitrogen	N	1.5	0.70		W
Oxygen	O	1.40	0.66	1.40 (−2)	P
Phosphorus	P	1.9	1.10		P
Potassium	K	2.25		1.33 (+1)	W
Rubidium	Rb	2.44		1.48 (+1)	W, P
Selenium	Se	2.0	1.16	1.98 (−2)	P
Silicon	Si		1.17		P
Silver	Ag	1.44		1.26 (+1)	W
Sodium	Na	1.86		0.95 (+1)	P
Strontium	Sr	2.13		1.13 (+2)	W
Sulfur	S	1.85	1.04	1.84 (−2)	P
Tin	Sn	1.40		1.02 (+2)	W
Titanium	Ti	1.46		0.76 (+2)	W
Tungsten	W	1.36			W
Zinc	Zn	1.33		0.74 (+2)	W
Zirconium	Zr	1.60			W

Radius (cm × 10⁻⁸): Van der Waals or metallic, Co-valent, Ionic[a].

[a] Ionic radius for ion whose charge is shown in parentheses.
[b] W is Wyckoff, *Crystal Structures*, vol. I, Interscience, New York and London, 1948.
P is Pauling, *Nature of the Chemical Bond*, 3rd edition, Cornell University Press, Ithaca, New York, 1960.
Both letters indicate: 1st figure is source of first letter, 2nd of second letter.

CRYSTAL STRUCTURES OF THE ELEMENTS ACCORDING TO WYCKOFF

(Wyckoff, *Crystal Structures*, Vol. I, Interscience Publishers, New York and London, 1948)

Close-packed metals			Body-centered cubic metals	
Cubic		Hexagonal		
Element	Unit cell dimension (cm $\times$ 10^{-8})	Element	Element	Unit cell dimension (cm $\times$ 10^{-8})
Ag	4.08	Be	Ba	5.01
Al	4.04	Cd	Cr	2.88
Au	4.07	Co	Cs	6.05
Ca	5.57	Cr	a-Fe	2.86
Ce	5.14	Mg	K	5.20
Co	3.55	Ni	Li	3.50
Cu	3.61	Ti	Na	4.28
Ni	3.52	Zn	Rb	5.62
Pb	4.94	Zr	Ti	3.32
Pt	3.92		V	3.03
Sr	6.05		W	3.16
Th	5.04			

Elements with the diamond structure

Element	Unit cell dimension (cm $\times$ 10^{-8})
C (diamond)	3.56
Si	5.42
Ge	5.62
Sn (gray)	6.46

CRYSTAL STRUCTURES OF COMPOUNDS MX ACCORDING TO WYCKOFF
(Wyckoff, *Crystal Structures*, Vol. I, Interscience Publishers, New York and London, 1948)

COMPOUNDS MX (M—metal, X—nonmetal) (Listings not complete)

NaCl arrangement:

Formula	Unit cell dimension	M–X distance	Formula	Unit cell dimension	M–X distance
LiH	4.09	2.04	AgF	4.92	2.46
LiF	4.02	2.01	AgCl	5.55	2.77
LiCl	5.13	2.57	AgBr	5.77	2.88
LiBr	5.49	2.75	MgO	4.20	2.10
LiI	6.00	3.00	MgS	5.19	2.59
NaF	4.62	2.31	CaO	4.80	2.40
NaCl	5.63	2.81	CaS	5.68	2.84
NaBr	5.96	2.98	SrO	5.14	2.57
NaI	6.46	3.23	SrS	5.87	2.93
KF	5.34	2.67	BaO	5.52	2.76
KCl	6.28	3.14	BaS	6.35	3.17
KBr	6.58	3.29	TiO	4.24	2.12
KI	7.05	3.52	MnO	4.43	2.22
RbF	5.64	2.82	MnS	5.21	2.61
RbCl	6.54	3.27	FeO	4.33	2.16
RbBr	6.85	3.43	NiO	4.17	2.08
RbI	7.33	3.66	CdO	4.69	2.34
CsF	6.01	3.00	PbS	5.92	2.96

CsCl arrangement:

Formula	Unit cell dimension	M–X distance
NH_4Cl	3.87	3.35
NH_4Br	4.05	3.50
CsCl	4.11	3.57
CsBr	4.29	3.71
CsI	4.56	3.95

ZnS arrangement:

Formula	Unit cell dimension	M–X distance
CuF	4.26	1.85
CuCl	5.41	2.35
CuBr	5.68	2.46
CuI	6.04	2.62
AgI	6.47	2.80
ZnS	5.41	2.36
CdS	5.82	2.52
HgS	5.84	2.53
SiC	4.35	1.89
BeS	4.85	2.10
AlP	5.42	2.35

All above dimensions in cm $\times$ 10^{-8}.

CRYSTAL STRUCTURES OF COMPOUNDS MX$_2$
ACCORDING TO WYCKOFF

(Wyckoff, *Crystal Structures*, Vol. I, Interscience Publishers, New York and London, 1948)

COMPOUNDS MX$_2$ (M—metal, X—nonmetal) (Listings not complete)

Fluorite arrangement (dimensions in cm × 10^{-8})		Anti-fluorite[a] arrangement		Rutile arrangement[b] (dimensions in cm × 10^{-8})		
Crystal	Unit cell dimension	Crystal	Unit cell dimension	Crystal	Unit cell dimensions	
					a	b
BaF$_2$	6.19	K$_2$O	6.44	CoF$_2$	4.69	3.19
CaF$_2$	5.45	K$_2$S	7.39	FeF$_2$	4.67	3.30
CdF$_2$	5.40	K$_2$Se	7.68	MgF$_2$	4.66	3.08
CuF$_2$	5.41	K$_2$Te	8.15	MnF$_2$	4.87	3.28
HgF$_2$	5.54	Li$_2$O	4.62	NiF$_2$	4.71	3.12
PbF$_2$	5.94	Li$_2$S	5.71	ZnF$_2$	4.72	3.13
SrCl$_2$	7.00	Li$_2$Se	6.01	CrO$_2$	4.41	2.86
SrF$_2$	5.78	Li$_2$Te	6.50	GeO$_2$	4.39	2.86
CeO$_2$	5.41	Na$_2$O	5.55	MnO$_2$	4.44	2.89
HfO$_2$	5.12	Na$_2$S	6.53	PbO$_2$	4.93	3.37
ThO$_2$	5.59	Na$_2$Se	6.81	SnO$_2$	4.72	3.16
UO$_2$	5.47	Na$_2$Te	7.31	TiO$_2$	4.49	2.89
PbO$_2$	5.58	Rb$_2$O	6.74	VO$_2$	4.54	2.88
ZrO$_2$	5.07	Rb$_2$S	7.65	WO$_2$	4.86	3.77
		Be$_2$C	4.33			

[a] In this arrangement, the metal atoms occupy the positions held by non-metal atoms in normal fluorite, and vice versa.

[b] See diagram, p. 208.

(b) This arrangement, diagrammed below, is one whose model was NOT described. The compound TiO_2, known as rutile, has each O surrounded by 3 Ti atoms in a triangular planar arrangement, while each Ti is surrounded octahedrally by six oxygen atoms.

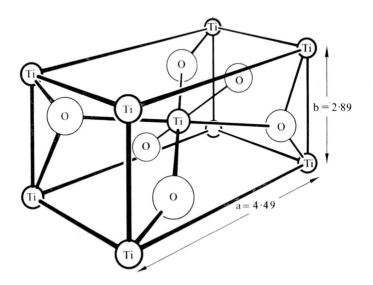

REFERENCES

General Chemistry
> SIENKO and PLANE, *Chemistry*, 2nd edition, McGraw-Hill, New York and London, 1961.
>
> SISLER, VANDERWERF, DAVIDSON, *General Chemistry*, Macmillan, New York, 1949.

> *More advanced*
>> MOELLER, *Inorganic Chemistry*, Wiley (N.Y.) and Chapman & Hall (London), 1952.

Structural Chemistry
> PAULING, *Nature of the Chemical Bond*, 3rd edition, Cornell University Press, Ithaca, New York, 1960.
>
> WELLS, *Structural Inorganic Chemistry*, 2nd edition, Oxford University Press, London and New York, 1950.
>
> WELLS, *The Third Dimension in Chemistry*, Oxford University Press, London and New York, 1956.

Crystallography and X-ray Analysis
> BRAGG and BRAGG, *The Crystalline State*, G. Bell & Sons, London, 1955.
>
> EVANS, *Crystal Chemistry*, Cambridge University Press, London and New York, 1946.
>
> LONSDALE, *Crystals and X-rays*, D. Van Nostrand, New York, 1949.

Models: Construction and Use of
> SANDERSON, *Chemical Periodicity*, Reinhold (N.Y.) and Chapman & Hall (London), 1960.
>
> SANDERSON, *Teaching Chemistry With Models*, D. Van Nostrand, New York, 1962.

Specific Reference Handbooks
> *Miscellaneous*
>> LANGE, *Handbook of Chemistry*, latest ed., Handbook Publishers, Sandusky, Ohio.

> *Atomic sizes and arrangements*
>> THE CHEMICAL SOCIETY, *Interatomic Distances*, Chemical Society, London, 1958.
>>
>> WYCKOFF, *Crystal Structures*, Vol. I, Interscience, New York and London, 1948. (Loose-leaf additions have been made since then.)

209

Sources of Supply for Model-making Materials and Supplies

Styrofoam Spheres (and sheet styrofoam)

Elford Plastics, Wood Street, Elland, Yorks., and from various laboratory suppliers—e.g. Griffin & George Limited, Ealing Road, Alperton, Wembley, Middlesex.

Plasteel Corp., 26970 Princeton, Inkster, Michigan 48141, U.S.A.

Star Band Company, Broad and Commerce Sts., Portsmouth, Virginia, U.S.A.

Probably not a direct source: Dow Chemical Co., Midland, Michigan, U.S.A.

Sheet styrofoam is a common packing material and may be obtained from builders merchants, etc.

Dowel Rod

Drugstore (ask for Peerless Applicators, in boxes of 72 dozen) in U.S.A. Local lumber dealer.

Jig Punches

Local welding supplier (ask for bronze $\frac{1}{16}$-in. welding rod), or hardware store (ask for $\frac{1}{16}$-in. stiff wire).

Paint

Water-soluble, latex-based paints, and second-coat enamels from local paint store.

Miscellaneous Materials: from local hardware dealer.

(Water soluble glue, round toothpicks, sand- and emery paper, wire-cutting pliers.)

Cardboard for Jigs

Use empty cardboard cartons.

INDEX

211

5420